Summary of the geological time scale, based on International Union of Geological Sceinces Global Stratigraphic Chart (IUGS 1989), together with the main phases of glaciation in Earth's history. Names in italics are based on those regions where the evidence is best displayed.

Eon	Era	Period	Glacial phases		Age (Ma)
Phanerozoic	Cenozoic	Quaternary	Antarctic	Northern Hemisphere	2
		Neogene			23
		Palaeogene			65
	Mesozoic	Cretaceous			135
		Jurassic			205
		Triassic			250
	Palaeozoic	Permian	Karoo		290
		Carboniferous			355
		Devonian	Niger		410
		Silurian	Saharan		438
		Ordovician			510
		Cambrian	Late Sinian		570
Proterozoic	Neoproterozoic		Varanger/Sturtian/Lower Congo		1000
	Mesoproterozoic				1600
	Palaeoproterozoic		Huronian		2500
Archaean					

Frontispiece: vertical air photograph of glaciers at the head of Engelskbukta ("English Bay") in NW Spitsbergen. This demonstrates the complexities of the glacial environment. Small cirque glaciers with arêtes between, and moraines near their snouts occur at the top of the picture. Two large valley glaciers (Uvêrsbreen; mid-right, and Comfortlessbreen, lower right) approach the bay from the right, both having contorted medial moraines and complexes of push moraines close to, and beyond, the ice limits. Large areas of braided outwash, both active and abandoned, are evident in the middle of the picture, the main channel complex producing a prominent sediment plume as the glacial meltwaters enter the sea. Comfortlessbreen ends partially in tidewater as a prominent calving ice cliff and, in front of it, beach ridges or spits of sand and gravel enclose muddy lagoonal areas. The field of view is approximately 8 km from top to bottom (photograph courtesy of Norsk Polarinstitutt, RC10, dated 7 August 1977).

Glacial Environments

Michael J. Hambrey
Liverpool John Moores University

UBC PRESS / VANCOUVER

First published in 1994 by UCL Press Limited
The name of University College London (UCL) is a registered trade mark
used by UCL Press with the consent of the owner.

Published simultaneously in Canada by
UBC Press, University of British Columbia,
6344 Memorial Road, Vancouver, British Columbia V6T 1Z2

Canadian Cataloguing-in-Publication Data

Hambrey, M. J.
  Glacial environments

  Includes bibliographical references and index.
  ISBN 0-7748-0509-9 (bound), –ISBN 0-7748-0510-2 (pbk.)

  1. Glaciers. I. Title.
GB2403.2.H35 1994   551.3'12   C94-910476-0

Typeset in Times New Roman and Optima.
Printed and bound by
Butler & Tanner Ltd., Frome, England.

Contents

Preface

Glacial environments are scenically and scientifically among the most exciting and complex on Earth. Apart from phenomena associated directly with moving or stagnating glaciers, fluvial, aeolian, lacustrine and marine processes frequently interact with ice and its deposits. Glacial environments therefore possess a wide variety of landforms and sediment associations.

Some 10% of the Earth's land surface today is covered by ice, a figure that exceeded 30% during the Quaternary glaciations of the past two million years. In earlier geological history, the Earth underwent glaciations of continent-wide extent on several occasions, some of them even more intense than those of the Quaternary period. Yet these earlier glaciations have received little attention compared with other global climatic events. Furthermore, the study of ancient glacial sequences has, until recently, been relatively simplistic, and frequently revealed an unawareness of the complexities of glacial processes.

The principal aims of this book are as follows: (a) to examine the processes associated with contemporary ice masses, especially those that can be observed or monitored in the field; (b) to emphasize the range and character of erosional and depositional landforms present at contemporary glacier margins, and those left by the Pleistocene ice sheets; (c) to draw examples from both Quaternary and ancient glacial periods, thereby providing a link between what we see at the surface and the rock record; and (d) to give equal emphasis to the glaciomarine environment which has been neglected in most textbooks.

The book deliberately avoids the use of most mathematical formulae since the emphasis here is on what can be seen and interpreted on the ground, or otherwise sampled. There are other excellent works, noted throughout the text, which provide rigorous mathematical treatment for those who desire it, but here the subject matter has been chosen to be accessible to those without advanced mathematical training. It is anticipated that the book will be of value to undergraduates studying physical geography, geology, environmental science and countryside management, as well as to secondary/high school teachers and those who wish to understand better the processes that have shaped our lands. No previous knowledge of the subject is assumed, other than an elementary knowledge of geological concepts.

By historical accident, Quaternary studies and glacial processes have been largely neglected in the traditional British geology departments. Rather, they commonly have formed a core element in physical geography courses, but here there has been a tendency to avoid rigorous application of sedimentological principles. In contrast, in North America, there has always been a much closer affinity between geomorphology and geology. The strategy adopted in this book is to integrate the approaches of the two disciplines as far as possible.

With the "greening" of earth sciences, and the development of employment opportunities in such areas as waste disposal, water management and the extraction of aggregates, earth scientists need to know more about the superficial glacial deposits that cover so much of the populated areas of the Northern Hemisphere. Ancient glacial deposits are also of economic importance since, in some regions, they provide petroleum plays, as in the Permo-Carboniferous and late Precambrian of the Middle East and South America. Furthermore, recent research, especially deep drilling on the continental shelves, has demonstrated the volumetric importance of glacial sediments.

In a relatively small book such as this, it has not been possible to reference relevant literature thoroughly. Instead, I have cited key references together with those that provide particularly good examples of the phenomena under review. Nor has it been possible to describe all components of the glacier system as one might wish; for example, glaciofluvial and glaciolacustrine environments are treated only lightly to allow for more thorough treatment of less-readily obtainable information on the glaciomarine environment. However, if the diversity and fascination of glacial environments becomes apparent through reading this book, it will have served its purpose.

MICHAEL J. HAMBREY
Liverpool, 1994

"O ye ice and snow, bless ye the Lord, praise him and magnify him forever."
Benedicite Omnia Opera
Book of Common Prayer

Acknowledgements

This book would not have been possible without the efforts of many colleagues throughout the world in promoting the subject. Foremost among them are W. H. Theakstone who was my mentor when I first began working in glacial environments; A. G. Milnes who convinced me of the strong linkage between glaciology and structural geology; and W. B. Harland who led me into investigations of ancient glacial sediments in areas where modern ones were being formed.

I am indebted to leaders of various expeditions for opportunities to work in glacial environments: P. Worsley in northern Norway; the late F. Müller in the Canadian Arctic; W. B. Harland in Svalbard; N. Henriksen in East Greenland; P. J. Barrett, W. U. Erhmann, D. K. Fütterer and G. Kuhn in Antarctica. The leaders of various conference- or workshop-based field excursions have also played an important rôle in providing a forum for stimulating discussion and opportunities to see glacial phenomena in many different settings.

Several colleagues, with whom I have undertaken joint research, have, over the years, provided a sounding board for ideas about glacial environments, notably J. A. Dowdeswell, I. J. Fairchild, A. C. M. Moncrieff and M. J. Sharp. In particular, I am grateful to the following for reviewing the whole of this volume: I. J. Fairchild, P. L. Gibbard and D. Huddart. I wish to thank R. Jones of UCL Press for maintaining the pressure to complete this work, and for his advice, and K. Williams, also of UCL Press.

Lastly, I acknowledge the help given by several funding agencies and institutions who have supported my work in regions influenced by glaciers today and in the past: the University of Manchester, the Swiss Federal Institute of Technology (Zürich), the University of Cambridge (especially the Scott Polar Research Institute, where the bulk of this book was written), Liverpool John Moores University, the Alfred Wegener Institute for Polar and Marine Research (Bremerhaven), Victoria University of Wellington, the Natural Environment Research Council, the Royal Society, the Geological Survey of Greenland and the Cambridge Arctic Shelf Programme.

For permission to reproduce figures for this volume, the various publishers and authors are thanked as noted in the appropriate places in the text. All photographs were taken by the author, except where otherwise stated.

1 Introduction

1.1 Historical background

1.1.1 Discovery of "ice ages"

Today it is common knowledge that the Earth once experienced an ice age, during which ice sheets spread over large parts of Eurasia and North America. We are all familiar with the sensationalist press reports that have inevitably arisen after exceptionally severe blizzards, debating whether a new ice age has come upon us, even if these have now been replaced by equally alarmist statements that human-induced global warming will cause catastrophic melting of the remaining polar ice sheets. However, when the concept of widespread glaciation in the past was first mooted in the early 19th century, it met with much opposition. The battles and personality conflicts that arose as a result throughout the first half of the 19th century provide a fascinating insight into how geology evolved into a science and overcame rigid prejudices that were based on a misunderstanding of Old Testament accounts of natural disasters, such as Noah's flood (see more detailed accounts in Garwood 1932, Imbrie & Imbrie 1979, Mills 1983).

The chief protagonist of the ice-age theory in the early 1800s was the influential president of the Swiss Society of Natural Sciences, Louis Agassiz, and it is he who came to be regarded as the "Father of Ice Ages". However, the idea that ice had once covered more of the Earth was not new. Agassiz was not the first to believe that glaciers had been more extensive; indeed, for many years he was sceptical. It was others, not all of them scientists, who documented the evidence for more widespread ice. Probably the first to do this was a Swiss minister, Kuhn, who, in 1787, interpreted local erratic boulders below the glaciers near Grindelwald as evidence of ice having been more extensive. In 1795, the leading geologist of the day, the Scot Hutton, published his *Theory of the Earth* in which he described how ice had transported great boulders of granite into the Jura Mountains. In 1815 a Swiss mountaineer and hunter, Perraudin, argued that glaciers extended much further down, and filled, the Val de Bagnes in the Alps. He expressed his views to a sceptical Charpentier, who later became an ardent advocate of the glacial theory. Three years later, Perraudin tried to persuade the Swiss engineer, Venetz, but he too had doubts about the validity of the idea. However, slowly Venetz began to accept the hypothesis and by 1829 was able to argue from the distribution of moraine and erratics that glaciers once covered the Swiss plain, the Jura and other parts of Europe. Already, in 1824, Esmark had argued that the glaciers of Norway had once been more extensive.

None of these men made the intellectual leap to conceive of a period of widespread global cooling and ice-sheet development. It was left to the famous German poet Goethe to promote the idea of an ice age ("Eiszeit"). Taking note of the findings by scientists of erratics on the North German Plain, Goethe developed his ice-age concept around 1823 in a novel, *Wilhelm Meister* (Cameron 1965).

Meanwhile, Charpentier, at last converted to the idea of more extensive ice, accepted Venetz's interpretation, and thereafter began to assemble a mass of evidence in its favour. Resistance was strong, since it was generally held at the time that the large erratics were deposited by Noah's flood. By 1833, many scientists had come to accept the view of the leading British geologist of the day, Lyell, that the boulders had been rafted by icebergs. This theory had originated with the German mathematician, Wrede, in 1804. The iceberg-rafting or "drift" theory, which apparently explained far-travelled erratic boulders, was expounded by Lyell in perhaps the most influential geology textbook ever written, *The principles of geology*, published in 1833. This explanation embraced conveniently the Flood theory, by providing a mechanism by which sediment, and especially large boulders, could have been transported – hence the term "drift" for these deposits. Lyell was strongly supported by Darwin who, in a series of papers, became a prominent advocate of the theory until his death in 1882.

Charpentier had as a friend Agassiz, who by now was one of Europe's leading scientists but, despite Charpentier's powers of persuasion, the young man was at first unable to accept the glacial theory. Eventually, during a field trip to Bex, Agassiz was won over, and for the first time the glacial theory had a strong, forceful and influential character to promote it. Both Agassiz and Charpentier had by now acknowledged Goethe's theory of a great ice age, although most scientists of the day ignored it, perhaps because of the poet's lack of scientific credentials. Unfortunately, Agassiz developed the glacial theory further, taking liberties with Charpentier's work, and promoting it beyond the available evidence. Thus, when Agassiz presented his work to the Swiss Society of Natural Sciences at Neuchâtel in 1837, he met with almost universal opposition. Nevertheless, he had a prominent ally in Germany at this time, Schimper, who provided him with much information about the former extent of glaciers in the Isar and Würm valley. Undaunted by the general opposition, Agassiz wrote up his work in the book *Etudes sur les glaciers* (*Studies on glaciers*), which was published in 1840, acknowledging the important work of his predecessors Venetz and Charpentier, but curiously not Schimper, and he set about trying to convince other scientists and the public at large. His belief in an ice age that caused great devastation and extinguished many animal species fitted in well with the philosophy of catastrophism that was prevalent in geology throughout the 18th and 19th centuries. The ice-age concept merely substituted one catastrophe for another, the Great Flood. One of the major proponents of the Flood theory was Buckland who, in addition to being a clergyman, was a Professor of Mineralogy and Geology at Oxford University, and so was well placed to explore the links between geology and religion. Buckland was also one of the first geologists to focus specifically on the accumulations of unconsolidated mud, sand and gravel, that covered much of the British Isles, and which were referred to as "Diluvium" by those who believed in the Flood. Buckland's impressive account of these deposits was published in 1823 and gained him immense respect.

Buckland meanwhile found it impossible to explain all the evidence in terms of a great flood with icebergs floating around. In particular, he wanted to know where all the water had come from and where it had gone. After hearing Agassiz promote his ideas in another meeting in 1838, in Germany, Buckland joined Agassiz on a trip to the Alps, but remained unconvinced that glaciers were responsible for drift deposits elsewhere. In 1840, Agassiz took on the British, reading a paper to the Geological Society of London "On the evidences of the glaciation of Great Britain and Ireland". This visit was portrayed in the satirical magazine *Punch* as "a sporting tour in the search of moorhens (moraines)". By now, Buckland had changed his mind and, after having shown Agassiz drift deposits around Scotland and northern England, became a strong advocate of the glacial theory himself. Within just a few months Buckland had converted Lyell to the theory, but, even with a trio of internationally renowned scientific heavyweights embracing it, wider opposition was not overcome immediately. Indeed, perhaps under the influence of Darwin, Lyell lapsed back into renewed support for the drift theory.

Although by 1841, Forbes (a Professor of Natural History at Edinburgh University, and himself a key figure in the development of glaciological concepts) was able to write to Agassiz "You have made all the geologists glacier-mad here . . . ", it was not for another 20 years that the majority of British geologists accepted the ice-age theory, following publication of classic papers by Jameson in 1862 and Geikie in 1863.

Various reasons have been given as to why Agassiz had so much difficulty in overcoming entrenched beliefs. Apart from the religious views, it proved difficult to explain the widespread "shelly drifts" around the coasts of northwest Europe. Furthermore, there was ignorance among geologists about glaciers themselves; it was not until 1852 that Greenland was discovered to have an ice sheet, and only towards the end of the century that Antarctica too was found to be covered by one. Agassiz himself did not help his cause, because in his enthusiasm he envisaged glaciers in places where the evidence was non-existent, such as the Mediterranean or the Amazon Basin.

In 1847 Agassiz moved to the USA, as Professor at Harvard University, and found that many American scientists had already accepted his theory. His arrival, however, did speed up its acceptance, and when Agassiz finally died in 1873, few scientists held out against it.

Following the establishment of an ice-age explanation for the widespread unconsolidated sediments, that ultimately were equated with the Quaternary period, it was only natural that geologists should seek for evidence of ice ages in the older rock record (Harland & Herod 1975). In 1855 Ramsey suggested that English Permian breccias were of glacial origin. Although he was incorrect in this, others began looking for Permian glacial sediments elsewhere, and by 1859 unequivocal deposits had been reported from India and Australia, and by 1870 from South Africa. In 1871 Precambrian glacial sediments were described from Scotland, and in 1891 from northernmost Norway where a striated pavement was also discovered. Evidence of older Precambrian and early Palaeozoic glaciations were found in the early 20th century.

Many discoveries of tillites were reported from around the world subsequently, but a phase of doubt, particularly with regard to the late Precambrian glaciation, entered the mind of many geologists as recently as the 1960s, perhaps because of the uncritical acceptance of many deposits as glacigenic at a time when the subdiscipline of sedimentology was revolutionizing the interpretation of sedimentary sequences. The few who maintained a pro-glacial stance for the Precambrian deposits (notably Harland 1964) have since been fully vindicated. A benchmark contribution that provided a thorough, objective account of the Precambrian glacial deposits of Scotland (Spencer 1971), set a standard which others have emulated with considerable success.

While the evidence for ancient ice ages was gradually being built up, the extent and number of glaciations in the Quaternary period were being documented. Two schemes, in particular, became widely accepted around the turn of the century. In the Mid-West of North America, Chamberlain and Leverett mapped four sheets of glacial drift, each representing a distinct "ice age". In the region to the north of the European Alps, Penk and Brückner similarly derived four ice ages, but by associating gravel terraces at progressively lower levels with cold periods when deposition was rapid. Similar successions were subsequently derived for the Scandinavian and British ice sheets, and even New Zealand. Despite the realization that successive glaciations tend to destroy the evidence of earlier ones, at least on land, the chronologies derived for the Mid-West and Europe were accepted uncritically, and the two were inevitably correlated, despite the absence of dating evidence. These schemes survived intact until the deep-sea record began to yield a different story.

When the first deep-sea sediment cores, going well back into the Quaternary period, were obtained during the 1950s and onwards, oxygen-isotopic studies of planktonic foraminifera enabled palaeotemperatures and ice-volume changes to be determined. The cores revealed rather more glacial periods than the commonly accepted terrestrial record, and land geologists for a long while ignored the evidence. In the 1960s, with the development of magnetostratigraphy and the establishment of magnetic reversals, the much-needed method of dating Quaternary events had finally arrived. Sediment cores, analyzed by an international team of scientists on a project called CLIMAP, gradually began to yield a climatic record that matched remarkably well the temporal changes in solar radiation derived by Milankovich from astronomical variables. The number of ice ages during the Quaternary period thus proliferated, and the fourfold ice-age chronology from the land areas was finally shown to be more incomplete than complete.

Until the early 1970s, it was assumed by many Earth scientists that the period represented by ice ages was essentially equivalent to the Quaternary period. In 1972, the Deep Sea Drilling Project extracted long cores from the Antarctic continental shelf in the Ross Sea (Hayes et al. 1975). To many people's surprise, evidence of glaciation extending back 25 million years (Ma), to the late Oligocene Epoch was preserved in these cores. In the past 20 years, Antarctica has yielded further evidence of the antiquity of Cenozoic glaciation. A succession of New Zealand drilling operations in the Ross Sea culminated in a drill-hole in 1986 that showed that glacier ice was present at least as far back as earliest Oligocene time (36 Ma). Most recently (1987–8), the Ocean Drilling Program in Prydz Bay has confirmed the existence of a large ice sheet over East Antarctica dating back at least this far. However, none of these drill-holes penetrated the glacial/preglacial boundary, and the onset of Cenozoic glaciation remains to be determined.

Back in the Northern Hemisphere, the onset of ice-rafting, indicating the development of an ice sheet over Greenland reaching the coast, has been dated at 2.4 Ma, with indications of ice-rafting in Baffin Bay from a Canadian Arctic source another million years earlier. Furthermore, glaciers in Alaska have been active since about 10 Ma on the high mountains there. Thus, we can no longer equate the Quaternary period with the development of ice ages, only with intensification of the ice cover.

1.1.2 Development of understanding of the dynamics of glaciers

To some extent, the development of ideas concerning the manner in which ice masses themselves behave has taken place independently of the investigations of the products of glaciation, although there have always been some scientists who have taken an interest in both fields. The history of the development of glaciological ideas is just as interesting and full of conflicting views as the establishment of the ice-age theory, and the reader is referred to Paterson (1981) and Clarke (1987) for fuller accounts.

The earliest descriptions of glaciers, in Icelandic literature, date from the 11th century. However, it was not until several centuries later that glaciers were recorded as being able to flow. By then the Earth was experiencing what became known as the Little Ice Age, which peaked about 1750, and resulted in strong advances of glaciers in many parts of the world. These events are particularly well documented where glaciers in the Alps and Norway destroyed pastures and even property, as well as being responsible for several disasters. Hence, a strong scientific interest in glaciers developed, notably in Switzerland.

Prominent among the early pioneers was Scheuzer who, between 1706 and 1723, published several works of a geographical and scientific character. He took a special interest in legends of Alpine dragons, but also studied glaciers, proposing that water entered fissures in the ice, and on freezing expanded, causing the glacier to thrust forwards – his so-called "dilation" theory. Altmann in 1751 and Grüner in 1760 explained that gravity was the cause of glacier motion, but assumed that this was accomplished entirely by ice sliding over its bed. In the late 18th century, Bordier suggested that ice can flow by internal deformation, somewhat like a viscous fluid.

Although Grüner had noted that stones on the surface of one of the Grindelwald glaciers had advanced 50 paces in six years, the first systematic measurements of glacier flow were not undertaken in the Alps until the 1830s. Foremost among the early experimentalists, once again, was Agassiz, who showed that ice moves faster in the middle than at the sides. The British scientists Tyndall and Forbes, and the Swiss Hugi and De Saussure, also became heavily involved in establishing the dynamics of glaciers in France and Switzerland by measuring ice movement and documenting surface structures. The Mer de Glace in France (Fig. 1.1) and the Unteraargletscher in Switzerland (Fig. 1.2) became favourite haunts. Forbes, having already fallen out with Agassiz after undertaking joint work on the Unteraargletscher, became involved in a heated dispute with Tyndall about the nature of glacier flow. Forbes believed that flow was of a viscous nature and considered ice to have many similarities to a metamorphic rock. Tyndall thought that motion resulted from the formation of small fractures that

Figure 1.1 The Mer de Glace in the French Alps, one of the earliest sites at which studies of glacier flow were first undertaken. The curving light and dark arcs are "ogives" or Forbes bands, each pair representing a year's movement through the icefall in the background. They were first described by James Forbes in the mid-19th century.

Figure 1.2 The Unteraargletscher in the Bernese Oberland of Switzerland. The first documented case of ice movement was recorded on this glacier when, over a period of several years, Hugi followed the displacement down-glacier of a large block on the medial moraine on the right.

were subsequently healed by pressure melting and refreezing, an idea that became known as the "regelation theory". Tyndall died in 1893 after his wife had unwittingly administered to him a lethal dose of choral, and with him the regelation theory also died. Forbes, on the other hand was essentially correct in linking glacier flow to fluid mechanics, although he did underestimate the rôle of basal sliding. His viscous flow theory motivated much laboratory experimental work and field measurements. Thus, by the end of the century, the manner in which ice at the surface of a valley glacier flowed was well known. In 1897, Reid in North America recorded the character of velocity vectors, inclined slightly downwards in the snow accumulation area and upwards in the ice-ablation zone, elaborating this in a classic paper "The mechanics of glaciers". However, ice movement at depth posed a different problem. Even though around the turn of the century Blümcke and Hess, using stakes set into a Tyrölean glacier, found that ice at depth moved faster than at the surface, many scientists for decades afterwards believed that the reverse was true and invoked a mechanism called "extrusion flow". Demorest and Strieff-Becker were particularly forceful proponents of this idea. After the considerable progress achieved up to 1900 it is strange that such an idea should have taken a strong hold.

The extrusion flow theory was not laid to rest until the 1950s, following deep borehole measurements on the Jungfraufirn in Switzerland and later on various Alaskan and Canadian glaciers, together with laboratory experiments on ice, and the application of modern ideas of solid-state physics, which showed that ice deformed in a manner similar to other crystalline solids such as metals and rocks. The foundations for our present understanding of glacier deformation were laid by British physicists Glen and Nye in the 1950s and 1960s, together with Lliboutry of France and Weertman of the USA, with their work on glacier sliding. Now we know that glacier ice deforms in a manner similar to plastic substances, and also slides on its bed. The rôle of the physicists in glaciology was expressed by Paterson (1981) in these terms: "a mere handful of mathematical physicists, who may seldom set foot on a glacier, have contributed far more to the understanding of the subject than have a hundred measurers of ablation stakes or recorders of advances and recessions of glacier termini". Unfortunately, since then, some mathematicians and physicists have tended to shift away from reality, and have derived equations that are virtually untestable in the field. However, the increasingly important rôle that glacier modellers are playing is bringing the observers and theoreticians back together again.

Major advances in other aspects of glaciology have taken place through the 20th century, such as the measurement of snow/ice density, accumulation and heat balance, the examination of glacier hydrology, and the palaeoclimatic record in ice sheets from drill cores, not only in valley glaciers but in the polar ice sheets (for example, Koch and Wegener in Greenland in 1913, and Ahlmann in Svalbard and elsewhere between 1920 and 1940). From 1957 (the International Geophysical Year) the Antarctic ice sheet has been investigated from all angles. Remote sensing techniques, such as radio-echo sounding and satellite imagery, have revolutionized our understanding of the extent, thickness and character of the ice on that continent, as indeed elsewhere.

Developments in linking the sedimentary record to glaciological principles have been a long time in coming. Geologists have been slow to understand the complexities of glacier dynamics, while many physicists have tended to assume that mathematics can provide all the answers and so have disregarded the evidence offered visually by glaciers. There are of course exceptions, and of these Boulton has perhaps done more than most in explaining the development of glacial sedimentary sequences in glaciological terms, starting with a series of investigations at the margins of glaciers in Svalbard in the 1960s, and extending to the large Northern Hemisphere ice sheets in the past decade.

In any account of the development of the science of glaciology in the past 50 years one cannot ignore the rôle played by the International (formerly British) Glaciological Society, which, through its *Journal of Glaciology* and more recently the *Annals of Glaciology*, has provided the principal focus for glacier research in all its aspects, and encoured interdisciplinary approaches to the subject. A history of the Society and its rôle in the development of glaciology has been provided by Weertman (1987).

1.2 Terminology and classification of glacigenic sediments

Before beginning this discussion, it is necessary to define a few basic terms, and here the definitions of Dreimanis (1989) are broadly followed, although their use in the literature varies.
 - **Glacigenic sediment** (also **glacigene**, **glaciogenic**): "of glacial origin"; the term is used in a broad sense to embrace sediments with a greater or lesser component derived from glacier ice.
 - **Glacial debris**: material being transported by a glacier in contact with glacier ice.
 - **Glacial drift**: all rock material transported by glacier ice, all deposits made by glacier ice, and all deposits predominantly of glacial origin deposited in the sea from icebergs, or from glacial meltwater.
 - **Diamicton**: a non-sorted or poorly sorted unconsolidated terrigenous sediment that contains a wide range of particle sizes (modified from Flint 1960).
 - **Diamictite**: the lithified equivalent of diamicton, and **diamict** embraces both (Harland et al. 1966). These terms, together with diamicton, have no genetic connotations.

Most investigations of glacigenic sediments, whether contemporary, Quaternary or ancient, have tended to use genetic terms for which no universal agreement has been reached. As a result, much confusion has ensued concerning the origin of a particular sediment. In the past decade it has been increasingly recognized that a study of a glacigenic sequence should begin with an objective description, before attempting to classify the sediments genetically.

1.2.1 Terminology and non-genetic classification of poorly sorted sediments

A variety of terms have been used in the past, mainly with reference to lithified deposits, to describe sediments without assuming a glacial origin. **Diamictite** has gradually found greater favour for lithified sediments than the synonymous **mixtite**, whereas **tilloid** has been used for "till-like rocks" in a variety of conflicting ways, and is also falling out of favour.

For the purposes of field investigation, a textural classification of diamictite has been devised

8

by Moncrieff (1989) and a modified version is used here (Table 1.1). It is based on the proportions of sand and mud (as matrix), discernible using a hand lens or with the naked eye, against the proportion of gravel clasts. In Cenozoic glacigenic sediments, the biogenic component may make up a considerable proportion of the sediment. Following Ocean Drilling Program procedures (Barron et al. 1989), prefixes such as shelly and diatomaceous may be used where such components exceed 30%.

Table 1.1 Non-genetic classification of poorly sorted sediments, based on Moncrieff (1989), but with maximum proportion of gravel in diamict reduced from 80 to 50% for compatibility with the Ocean Drilling Program's definition of diamict and conglomerate/breccia (Barron et al. 1989). The term "diamict" embraces both diamicton and diamictite. "Mud", as used in this context, covers all fine sediment, i.e. mixtures of clay and silt.

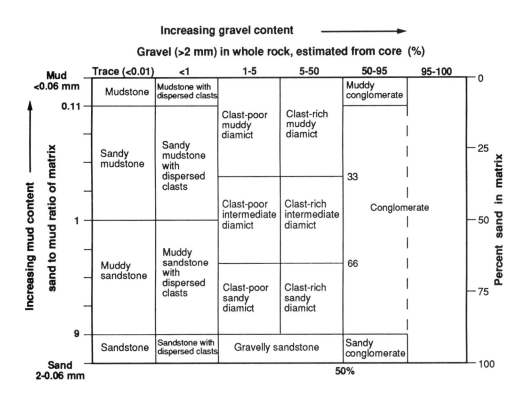

1.2.2 Classification of sorted sediments

Various other sedimentary types, not restricted solely to glacial environments, should also be mentioned: these include gravel, sand, silt and clay, with varying degrees of sorting; rhythmically bedded/laminated sediments of sand/silt/clay and non-cyclically deposited graded beds such as turbidites. The Wentworth (1922) scale of grain sizes is the most widely used, summarized as follows:

9

- boulder: >256 mm (< -8 phi units (ϕ))
- cobble: 64 to 256 mm (-6 to -8ϕ)
- pebble: 4 to 64 mm (-2 to -6ϕ)
- granule: 2 to 4 mm (-1 to -2ϕ)
- sand: 0.0625 to 2 mm (4 to -1ϕ)
- silt: 0.0039 to 0.0625 mm (8 to 4ϕ)
- clay: < 0.0039 mm ($<8\phi$)

"Gravel" embraces all classes >2 mm. The term "mud" is used in various ways; in this book, usage is consistent with many sedimentological studies in meaning a mixture of silt + clay.

1.2.3 Terminology of glacigenic sediments

The question of terminology of till and till-like deposits has received a thorough airing repeatedly over the past three decades (e.g. Hambrey & Harland 1981, with reference to pre-Pleistocene sediments; Dreimanis 1989, for Quaternary sediments). Here, the most important terms used in this and subsequent chapters are summarized.

For an unsorted deposit with a wide range of grain sizes deposited directly from glacier ice, whether on land or beneath a floating glacier, and not subsequently modified, the term **till** is applied. This term is an old Scottish word originally used by country-folk to describe "a kind of coarse obdurate land", the soil developed on the stony clay that covers much of northern Britain (Flint 1971: 148). It was adopted as a genetic term by Scottish geologists in the mid-19th century, and its use has spread across the English-speaking world and into other languages. Some authors restrict the term to material deposited on land, but as it is often difficult to distinguish the environment of deposition, the deposit could be labelled as till only if one were sure of the nature of the environment, thus restricting its usefulness. The term **boulder clay**, which has been used in the British Isles as a synonym of till, is no longer favoured by most glacial geologists. Moraine has also often been used as a synonym for till, but it is best to restrict this term to the landform.

The term for a lithified glacial deposit, **tillite**, historically has evolved separately and is not strictly equivalent. Many authors have used the term to embrace sediments containing a significant proportion of iceberg-rafted material. Other authors have been more restrictive, although few would restrict it solely to material known to have been deposited on land, but would include lithified till-like material deposited beneath a floating glacier. Here, the terms "till" and "tillite" are used to include sediments released directly from a glacier, whether on land or through a water column, that have not been subject to reworking, such as by currents or gravity flowage resulting in disaggregation.

Materials released by ice into the sea, whether by continuous rain-out beneath a floating mass of glacier ice, or sporadically from icebergs, even if the proportion is small, are collectively referred to as **glaciomarine sediments** (also referred to in the literature as glacial-marine, glaci(-)marine). Thus, the broad inclusive definition proposed by Andrews & Matsch (1983: 2) and Borns & Matsch (1989: 263) is adopted (but anglicized) here:

"Glaciomarine sediment includes a mixture of glacial detritus and marine sediment deposited more or less contemporaneously. The glacier component may be released directly from glaciers and ice shelves or delivered to the marine depositional site from those sources by gravity, moving fluids, or iceberg rafting. The marine component comprises mainly terrigenous and biogenic ('biogenous' in North America) sediments. Glaciomarine sediments vary laterally from ice-proximal diamicton, gravel and sand facies, to an intermediate pebbly silt and mud facies, to distal marine environments where the glacial imprint is seen in ice-rafted debris particles usually in the -1 to 4ϕ fraction".

1.2.4 Genetic classification of tills

Tills (and tillites) are more variable than any other sediment known by a single name (Flint 1971: 154; Goldthwait 1971: 5). Not surprisingly, therefore, the meaning of till varies from one investigator to another. Two extreme views have been published. The first, by Harland et al. (1966), used the term in a very broad sense to include any poorly sorted sediment that contains glacially transported material. By contrast, Lawson (1979) used a very restrictive definition of till by excluding those sediments that included components that were not *directly* glacially deposited, as well as those that showed any hint of reworking. Drewry (1986: 120) even stated that the genetic term "till" is no longer applicable and new nomenclature is necessary. One might think that such disparate views preclude any agreement about the definition of "till", let alone the development of a genetic classification. However, a considerable measure of agreement has been achieved by the International Quaternary Association's (INQUA) Commission on Genesis and Lithology of Quaternary Deposits. A comprehensive classification of tills has emerged that has satisfied the majority of INQUA correspondents (Dreimanis 1989). The INQUA classification represents the broadest consensus concerning glacial sediments at the present time and the terrestrial elements are adapted in this book (Table 1.2). The factors considered in the INQUA classification are primarily the formational and depositional processes, the general environment of deposition, and the position in relation to glacier ice (Table 1.2).

In the genetic depositional classification of till there are two main categories. **Primary tills** are formed mainly by direct release of debris from the glacier and are deposited by primary glacial processes, namely melt-out, lodgement, sublimation, or during deformation induced by the glacier. **Secondary tills** are the products of resedimentation of glacial debris that has already been deposited by the glacier, with little or no sorting by meltwater. Within these categories several varieties of till have been documented (Table 1.2), although these represent end members in a continuous spectrum of depositional types (Dreimanis 1989). **Melt-out till** is deposited by a slow release of glacial debris from ice that is not sliding or deforming internally. **Lodgement till** is deposited by plastering of glacial debris from the sliding base of a moving glacier by pressure melting and/or other mechanical processes. **Sublimation till** is till released by the sublimation (direct transition of ice to the vapour state) of debris-rich ice (Shaw 1989). It requires long-term extremes of cold and aridity for its formation, so its development is restricted to Antarctica. **Deformation till** comprises weak rock or uncon-

Table 1.2 Genetic classification of till in terrestrial settings (adapted from the INQUA classification, Dreimanis 1989: Table 11). The vertical columns are independent of each other and no correlation horizontally is implied. Not all combinations are feasible.

Release of glacial debris and its deposition or redeposition			Depositional genetic varieties of till		
I Environment	II Position	III Process	IV By environment	V By position	VI By process
Glacio-terrestrial	Ice-marginal frontal lateral	*A. Primary* Melting out Lodgement Sublimation Squeeze flow Subsole drag	Terrestrial non-aquatic till	Ice-marginal till Supraglacial till Subglacial till	*A. Primary till* Melt-out till Lodgement till Sublimation till Deformation till Squeeze flow till
	Supraglacial				
	Subglacial	*B. Secondary* Gravity flow Slumping			*B. Secondary till* Flow till
	Substratum	Sliding and rolling			

solidated sediment that has been detached by the glacier from its source, the primary sedimentary structures distorted or destroyed, and some foreign material admixed (Elson 1989). This term suffers from having various other meanings, not recognized by INQUA, so it should be used with care. **Squeeze flow till** is the result of squeezing or pressing of till by the weight, or movement, of glacier ice. **Flow till** may be derived from any glacial debris upon its release from glacier ice or from a freshly deposited till, in direct association with glacier ice. Redeposition is accomplished by gravitational slope processes, mainly by gravity-flow, and it may take place ice-marginally, supraglacially or subglacially, and subaerially or subaquatically.

These process terms may conveniently be used in combination with the terms for position, e.g. "supraglacial melt-out till". Many other terms for till and combinations have been used, of which Dreimanis (1989) has provided a comprehensive review.

Recognition of lodgement till, melt-out till and flow till has long been a matter for debate, and the most useful criteria are tabulated in detail by Dreimanis (1989: appendix D). Genetic terms for lithified sediments may have the suffix "-ite".

1.2.5 Genetic classification of glaciomarine sediments

In terms of preservation potential and volume, glaciomarine sediments are vastly more important than terrestrial glacial sediments, yet before the late 1970s little was known about the contemporary environment, and sediment classifications were not based on direct observations of the processes (e.g. Harland et al. 1966). Since then, however, glaciomarine environments have received considerable attention, notably those in Antarctica, Svalbard and Alaska, and several simple classifications have been adopted. For example, a genetic classification arising from wide-ranging American studies on the Antarctic continental shelf includes the following three main categories (Anderson et al. 1980, 1983):

- **Basal till** – deposited on the shelf by grounded glaciers (therefore better grouped with terrestrial sediments);
- **Compound glacial marine sediment** – resulting from a combination of ice-rafting (from icebergs and ice shelves) and normal marine sedimentation;
- **Residual glacial marine sediment** – the product of ice-rafting coupled with bottom current activity that is sufficiently strong to winnow silts and clays.

To these categories may be added a fourth – sediment gravity-flows – described by Wright & Anderson (1982) from Antarctica and by Miall (1983a) from the Early Proterozoic Gowganda Formation. From the perspective of position in relation to the ice margin of grounded tidewater glaciers in Svalbard, Boulton (1990) identified three main zones of sedimentation:

- **inner proximal zone** – sedimentation sufficiently high to inhibit benthic life and bioturbation is rare (0–7 km from the ice margin);
- **outer proximal zone** – where benthic life and bioturbation are common (7–60 km), and sedimentation is influenced by suspended and ice-rafted components;
- **distal zone** – outer fjord or outer shelf where suspended sediment concentrations are much less, upwelling of deep water along the continental margin occurs, and the sea bed is affected by waves and subject to reworking and erosion of finer materials.

Deep drilling, combined with seismic investigation on the Antarctic continental shelf has yielded important information about depositional processes in a temporal context (Barrett et al. 1989, Barron et al. 1989, 1991). This has led to a classification that is related to the proximity of the sediment source (Hambrey et al. 1989, 1991), and it depends on an assessment of the relative importance of rain-out deposition, ice-rafting and biogenic activity. In this case the "proximity" is less to do with the ice (much of which may not be delivering sediment to the sea) than with the main source of sediment, such as in ice streams. This classification, together with sediment gravity-flow is used here.

- **Waterlain till** – sediment that is released from floating basal glacier ice and accumulates on the sea bottom without being affected by winnowing processes. In character it resembles basal till, from which it may be difficult to distinguish without clast orientation measurements. Francis (1975) introduced the term "waterlain till" in preference to the semantically incorrect waterlaid till of Dreimanis (see Dreimanis 1979). Dreimanis inferred the same sort of depositional processes as summarized above, but texturally included both stratified and unstratified varieties of diamicton. This term has been adopted for Antarctic sediments (Hambrey et al. 1989, 1991), but Dreimanis (1989) has subsequently advocated abandoning the term. It is retained here, however.
- **Proximal glaciomarine sediment** – sediment that is composed principally of debris released from floating glacier ice and icebergs, and that has been affected by winnowing processes. A biogenic component in the form of shelly fauna and diatoms may also be present.
- **Distal glaciomarine sediment** – sediment that is principally of marine origin, such as suspended sediment and biogenic material, with a minor iceberg-rafted component (< 1% ice-rafted material).

A comprehensive provisional classification of glaciomarine processes and sediments was compiled for INQUA by Borns & Matsch (1989), following the general principles of the terrestrial classification. It draws attention to the wide range of ice-margin types and depositional processes, but does not define them or discuss how genetically different sediments may be distinguished, so its use at this stage is premature.

1.2.6 Genetic classification of glaciolacustrine sediments

Glaciolacustrine sediments embrace all material derived directly or indirectly from a glacier that is in contact with an enclosed standing-water body. Characteristically, they comprise sediments which have a direct glacial and iceberg component, and can be classified in the same way as those in the glaciomarine environment: waterlain till, proximal and distal glaciomarine sediment, and sediment gravity-flows. They also include deltaic deposits and **rhythmites**, notably **varves** or **varvites** (lithified). A varve may be defined as a sedimentary bed or lamina or sequence of laminae deposited in a body of still water, and representing one year's accumulation. A varve comprises a thin pair of graded glaciolacustrine layers, seasonally deposited (usually by meltwater streams) in a glacial lake or other body of still water in front of a glacier. The varve normally includes a lower summer layer consisting of relatively coarse-grained, light-coloured sediment (usually sand or silt) produced by rapid melting of ice in the warmer months, which grades upwards into a thinner winter layer, consisting of very fine-grained (clayey), often organic, dark sediment slowly deposited from suspension in quiet water while the streams were ice-bound. Counting and correlation of varves have been used to measure the ages of late Quaternary glacial deposits. The term was introduced by De Geer in 1912 (Swedish: *varv*). Although characteristic of glacial lakes, not all glacial lakes have varves, and not all varves are glacial. Non-glacial examples have been described, for example, in the papers published in Schlüchter (1979). Furthermore, some varve-like sediments form in fjord settings, but are not annual. Glaciolacustrine sediments may include material released directly from ice, by ice-rafting.

1.3 Facies analysis of glacigenic sediments

For most of the period in which glacigenic sediments have been studied, it has been customary, with regard to both Quaternary and pre-Quaternary sediments, to apply genetic terms (such as till) uncritically. This approach has often led to confusion, because tills and tillites are such varied deposits that they have meant different things to different people. As a result, many palaeo-environmental reconstructions have been based on inadequate data, or on data presented in a way that others cannot use.

It is now widely recognized that sequences of glacigenic sediments need to be examined objectively and, for this, facies analysis, as developed for other branches of sedimentology, is the most important approach. Some advocates of the facies approach have tended to be dismissive of other methods of analysis, but these too are needed if one is to gain a clear understanding of the mode of deposition and palaeo-environment.

1.3.1 The facies concept

The concept of **facies** has been used ever since it was recognized that features found in particular rock units were useful for interpreting the environment of deposition and for predicting the occurrence of mineral resources. (Reading 1978 gives a useful summary).

A sedimentary facies or **lithofacies** is a body of sediment or rock with specified characteristics, namely colour, bedding, geometry, texture, fossils, sedimentary structures and types of external contacts. The term "facies" has been used in many different senses, for example, in the strictly observational sense, in the genetic sense, and in an environmental sense. However, a facies should ideally be a distinctive rock that forms under certain conditions of sedimentation, reflecting a particular process or environment. Facies may be subdivided into subfacies or grouped into **facies associations** (Reading 1978), or considered on a regional scale in terms of **facies architecture**.

Here, glacial facies refers to the different sediment types one finds in a glacial environment, and which are interpreted as till, glaciofluvial, glaciolacustrine and glaciomarine deposits. Grouped together we have terrestrial glacial facies associations, glaciomarine facies associations, and so on.

1.3.2 Glacial sedimentary facies

Examination of contemporary glacial environments indicates that sedimentary facies are varied and related in an often complex manner. Identification of these facies in Quaternary sequences is relatively straightforward, but in detail, for example, it may be difficult to distinguish different types of glacigenic sediment. Glacial, fluvial, aeolian, marine, lacustrine and mass-flow processes account for the wide variety of facies present. In pre-Pleistocene sequences the frequent lack of three-dimensional exposure of strata often makes it difficult to determine the precise depositional environment and the degree of direct glacial influence in a particular facies. It is therefore important to undertake first a descriptive facies analysis, using the criteria listed in Table 1.3, applying non-genetic terms such as diamict (on/ite), and only then interpret them. There are many instances where authors have interpreted their sediments without providing adequate descriptions.

1.3.3 Interpretation of facies

Briefly, the principal descriptive facies are listed below, with comments on how each may be interpreted.

(a) Massive diamict with striated, predominantly angular to subrounded stones – basal till

Table 1.3 Principal descriptive criteria used in defining lithofacies in glacigenic sequences.

Lithology	Bedding characteristics	Bedding geometry	Sedimentary structures	Boundary relations
Diamict(on/ite)	Massive	Sheet	Grading: normal	Sharp
Gravel	Weakly stratified	Discontinuous	reverse	Gradational
Sand(stone)	Well stratified	Lensoid	coarse-tail	Disconformable
Mud(stone)	Laminated	Draped	Cross-bedding: tabular	Unconformable
	Rhythmic lamination	Prograding	trough	
	Wispy stratification		Lonestones (dropstones)	
	Inclined stratification		Clast supported	
			Matrix supported	
			Clast concentrations:	
			layers	
			pockets	
			Ripples	
			Scours	
			Load structures	
			Mottling (=bioturbation)	

deposited subglacially by melt-out or lodgement, or waterlain till deposited by steady rain-out of debris from the base of floating glacier ice without reworking, or till that has been subject to gravity flowage, either subaerially or **subaquatically**.

(b) Diamict with deformation structures (folds, thrusts, faults, convolutions) – deformation till formed by push or overriding by the glacier, or till that has undergone **gravity flowage**.

(c) Weakly stratified diamict with some microfossils and shells – deposited from the base of a floating glacier in a proximal glaciomarine or glaciolacustrine setting, with some reworking by bottom currents.

(d) Massive diamict with a greater or lesser degree of sorting (e.g. weak grading), sometimes underlain by soft sediment scour marks – subaquatic **slumping** of unstable till deposited on a slope (e.g. the continental slope).

(e) Massive breccia with angular stones, sparse fine material – supraglacial till derived from rockfall and deposited by melting of underlying ice.

(f) Laminated mud/silt with outsize stones and marine fossils – clastic marine sediments with ice-rafted **dropstones** mainly from calved icebergs.

(g) Rhythmites with regular sand/silt or clay couplets, occasionally with outsized stones – varves (lacustrine) or tidally controlled laminae (marine) with dropstones.

(h) Graded laminae (mud/sand) or beds (sand/gravel) of sporadic origin, for example, from **turbidity currents** in lakes or the sea.

(i) Stratified sands, cobbles and boulders, moderately well sorted and with subrounded to rounded stones, often with trough cross-bedding, ripples, mud-drapes – recycled till, transported and redeposited by running water (glaciofluvial) usually in a subglacial, proglacial and subaquatic environment; typically braided-stream facies or subaquatic outwash; alternatively, sand and gravel horizons may represent lag deposits, resulting from removal of fines from glacial sediments in the marine environment.

(j) Stratified and non-stratified, well sorted silts – aeolian deposits (**loess**) formed as a result of wind erosion and transport (**deflation**) of outwash plains and till-covered areas.

16

(k) Facies cutting across the bedding, such as diamict beds with upper reworked, sorted parts preserving wedge-shaped features, polygons, stripes and circles – periglacial phenomena generally indicating permafrost conditions.

(l) Any of the above facies with beds that have been down-faulted, folded, and locally depressed, and subsequently overlain by other sediments, are the result of differential melting of buried stagnant glacier ice.

1.3.4 Lithofacies coding

In descriptive facies analysis it is common to use shorthand notation, e.g. Dm for massive diamict. Eyles et al. (1983) established a formal lithofacies code in order to allow rapid description and visual appraisal of field sequences or drill-cores containing diamictons or diamictites, from which environmental interpretation can then be undertaken. This lithofacies approach had already been applied to braided-stream deposits by Miall (1977). Examples of its application are to the early Proterozoic Gowganda Formation of Ontario (Miall 1983a) and to the Cenozoic Yakataga Formation of Alaska (Eyles & Lagoe 1990).

The approach of Eyles et al. (1983) was strongly criticized by Dreimanis (1984), Karrow (1984) and Kemmis & Hallberg (1984) since, on its own, it was held to be too restrictive, and regarded other modes of till study (such as fabric and granulometric compositional analyses) as of secondary importance. The code also adds an interpretive letter, which deflects from the objective nature of the approach.

Independently of Eyles et al. (1983), Fairchild & Hambrey (1984) used a more simple facies abbreviation for shorthand descriptive notation of glacigenic lithofacies. They preferred this to be informal, each study being expected necessarily to generate its own scheme. It was argued that lithofacies analysis alone was insufficient to establish the precise mode of deposition of all glacigenic sediment, and at least some of the other methods outlined in §1.4 need to be employed. A flexible approach, involving fresh appraisal of each sequence, allows lithofacies to be defined to suit the sequence concerned. Complex codes are useful as abbreviations for illustrative purposes, but tend to hinder the conveyance of information to the non-specialist reader.

1.3.5 Facies associations and facies models

The next step to considering lithofacies individually is to examine how they relate to one another, namely, the study of facies associations. Some facies are mutually exclusive. For example, the terrestrially deposited massive till may be associated with aeolian siltstones fluvial sands and gravels, but not with widespread rhythmites containing dropstones and fossils. The variety of facies represented in a facies association reflects the advance and recession of glaciers in terrestrial, lacustrine, intertidal or marine environments, or a combination of them. Hypothetical facies associations for the main environments are given in Figure 1.3, and actual examples from the sedimentary record are described in the relevant chapters.

Having considered the origin of facies and their spatial arrangement, one is then in a position to develop a pictorial representation of the palaeo-environment and the processes operating therein, that is to develop a **sedimentary model**. Such models have been developed for a range of glacial settings, some of which are discussed later. Models require a certain amount of generalization of the processes, since the main point of them is to examine how they apply to similar settings elsewhere.

1.3.6 Stratigraphic (or facies) architecture

Glacial geologists have long been concerned to establish the large-scale, three-dimensional geometry of glacigenic facies and their relationships to one another, especially in the ancient record. This approach is now used to define what in the past few years has become known as stratigraphic or facies architecture. The concept was applied first to fluvial sequences, when it was realized that lithofacies logging provided only part of the environmental picture. Its application is particularly appropriate for thick, laterally extensive, glaciomarine sequences, now aided significantly by the availability of seismic profiling across present-day continental shelves. Major advances have been made concerning the development of high-latitude continental shelves under the influence of ice sheets by combining drilling and seismic surveying (see Cooper et al. 1991a,b and Hambrey et al. 1992 for reviews concerning Antarctica, and King, 1993, with reference to northern latitudes). In contrast, most terrestrial successions are normally of limited areal extent, so the architecture of glacial sequences on land can be conceived on a much smaller scale than in the marine environment.

The large-scale three-dimensional architecture of sedimentary sequences reflects the organization of sedimentary environments in space and time. Each sequence demonstrates a range of interacting processes which, if repeated, will reproduce characteristic lithofacies associations.

Stratigraphic architecture of glacigenic sequences is a response to the interactions between four related phenomena: (a) the geometry of the crust, (b) the spatial and temporal pattern of expansion and decay of the ice sheet, and its relation to the global glacio-eustatic cycle, (c) isostatic response of the crust, and (d) patterns of ocean circulation. Stratigraphic architecture provides a framework whereby major global questions can be addressed, e.g. the response of sea-level changes to glaciation. A stimulating review of such an approach has been provided by Boulton (1990).

1.3.7 Sequence stratigraphy

A recent development in the analysis of sedimentary basins has been the adoption of a concept that has been in use by the petroleum industry since the 1970s, namely **seismic stratigraphy**. Companies have had to base their investigations mainly on seismic records in which the ages and character of the sediments are, as often as not, unknown, at least until exploratory wells have been sunk. **Sequence stratigraphy** has been a subsequent development of this approach and now provides a means of obtaining a high-resolution time-stratigraphic framework in which to place lithological and subsurface (well-log and reflection seismic) data. Proponents of the technique have argued that it allows one to undertake global correlations, although others have disputed these claims. Sequence stratigraphy was originally developed for interpreting marine seismic sections, but many articles applying the technique to onshore areas have now been published. However, sequence stratigraphy has not yet been applied to many glacial sequences, so there is much potential both for understanding better the relationships of facies in space and time, and for establishing the links between facies and sea-level changes such as those determined globally by Haq et al. (1987).

Sequence stratigraphy is essentially the exercise of defining packages of strata that are bounded by unconformities. A **sequence** is defined as "a stratigraphic unit composed of a relatively conformable succession of genetically related strata, and bounded at its top and base by unconformities or the correlative conformities, called **sequence boundaries**". The development of **sequence boundaries** is related to eustatic sea-level changes and has allowed the development of global sea-level curves, often referred to as "Vail-curves" after an Exxon petroleum geologist who first developed the concept. These curves have been refined over the years (e.g. in Haq et al. 1987). Developed principally for Mesozoic–Cenozoic sedimentary basins, the curves were explained in terms of changes in global ice volume, although for the Mesozoic Era at least the presence of ice sheets has not been substantiated. The reader is referred to Wilson (1991) for a helpful introduction to sequence stratigraphy.

1.4 Other methods of analysis of glacigenic sediments

1.4.1 Grain-size distribution

Diamictons of glacial origin have a broad range of grain sizes, with representatives in all classes, together with relatively subdued peaks or none at all. However, examination of the grain-size distribution is helpful in determining: (a) the nature of the source material, (b) how it becomes modified during glacier flow and when it is deposited, and (c) the extent to which it is modified following deposition. Examination of grain size can take place on all scales. For example, it is useful to estimate the proportion of material of gravel size (> 2 mm); this can be achieved visually using clast density charts. More common are analyses of the sand and finer fractions using a variety of methods ranging from basic sieving to the use of a sophisticated instrument such as the SediGraph, which uses X-rays to obtain the grain-size distribution.

Grain-size analysis has proved particularly useful in examining the processes that modify till after it has been deposited in the glaciomarine environment (e.g. Anderson & Molnia 1989, Barrett 1989a), where winnowing processes operate. With regard to terrestrial tills, individual sheets are remarkably uniform, but they may vary greatly from one to another (Sladen & Wrigley 1983). The regional variation between till sheets is related particularly to rock type and to the incorporation of pre-existing sediments by the glacier (Sladen & Wrigley 1983),

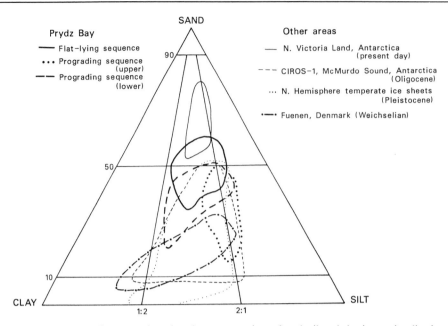

Figure 1.3 Ternary diagram to show the relevant proportions of sand, silt and clay in massive diamicts from different areas. Note that contemporary diamicts from Antarctica (Victoria Land) are sandy, whereas the Pleistocene examples from the Northern Hemisphere are more clay/silt-rich. The Prydz Bay and CIROS-1 samples are from older glacial sediments (mainly Oligocene) recovered in drill-holes on the continental shelf (From Hambrey et al. 1991; with permission of the Ocean Drilling Programme).

or even to the thermal regime of the ice mass (Barrett 1986). Figure 1.3 illustrates the grain-size distribution of various glacigenic sediments in a ternary diagram, including both terrestrial and continental shelf deposits from different climatic regimes.

1.4.2 Clast shape

A variety of parameters have been used for the analysis of particle shape in rocks (reviewed by Barrett 1980), and it is often difficult to compare one study with another. Perhaps the most useful for glacigenic sediments, because it links the sediments to the ice being transported, is the method presented Krumbein (1941) as applied by Boulton (1978). In this approach, Krumbein sphericity (based on ratios between the long, short and intermediate axes) is plotted against Powers roundness, estimated visually from shape charts. The shape charts also define fields of four different shapes: discs, spheres, blades and rods (Zingg shape) which also aid the comparison of different glacial sediments. Boulton found distinct, but overlapping fields for supraglacial (rockfall debris), basal till and lodgement till in Spitsbergen and Iceland. This study has formed the basis for comparison with the glaciomarine environment of Antarctica (Domack et al. 1980, Kuhn et al. 1993), and Baffin Island (Dowdeswell 1986), and with Precambrian glacial sediments from Svalbard (Dowdeswell et al. 1985; Fig. 1.4). In most studies, 50 clasts have been measured at each site.

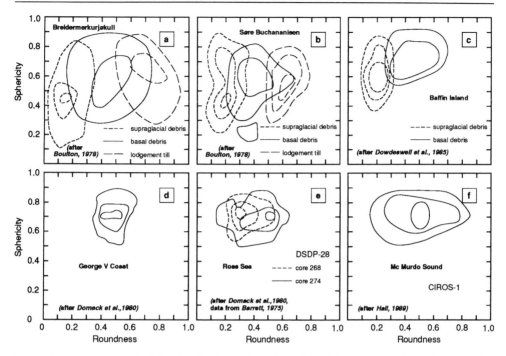

Figure 1.4 Roundness–sphericity plots for Antarctic continental shelf clasts, compared with those denoting specific glacial transport paths from Boulton (1978). The Icelandic glacier is temperate; the Svalbard glacier is slightly cold and a little more analogous to the Antarctic glaciers (from Kuhn et al. 1993; data from various sources).

1.4.3 Clast fabric

Measurement of the orientation of the long axis of stones is one of the oldest and most widely used techniques employed in the investigation of diamicts, and it has been extended to the smaller grains that are the only ones in sufficient numbers in cores. By comparison with glacial striations or flutes, it has been shown that diamicts commonly display a preferred orientation parallel to ice-flow, hence they are useful in establishing regional patterns of ice movement. However, transverse orientations are also possible, such as in an end moraine, or the fabric may be disrupted by post-depositional flowage of the sediment.

According to the level of sophistication required, one of two methods may be used. The simplest is to measure the long-axis orientation projected on to the horizontal plane. Data are grouped into convenient classes, e.g. 10°, 20°, 30°, and plotted as a rose diagram. A simple statistical check using the chi-squared test will establish the strength of the fabric at various confidence levels. This method is especially useful in material (e.g. cores) that does not lend itself to disaggregation but only permits one to take measurements on an exposed horizontal surface.

The second method, which is preferable, is to measure the plunge of the stone axis as well as its orientation, to give the three-dimensional view (Fig. 1.5). As with structural geological data, clast fabric data may be plotted on a stereographic projection, usually a Schmidt/

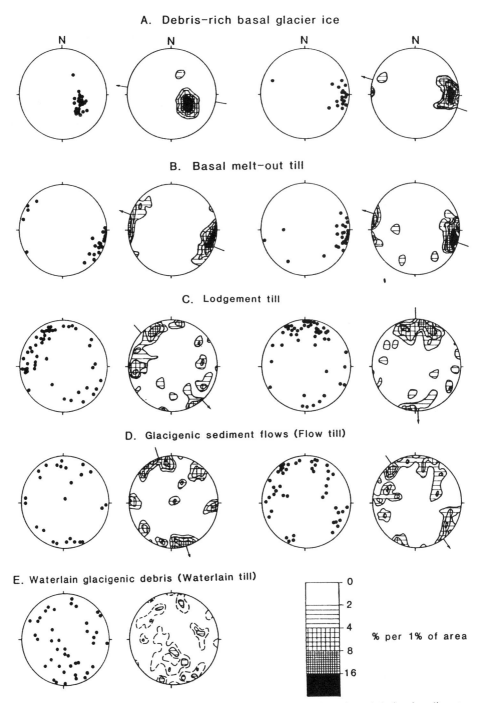

Figure 1.5 Three-dimensional clast-orientation data from basal ice and a variety of glacigenic sediments, plotted on the lower hemisphere of a Lambert/Schmidt equal-area projection. The contoured and shaded diagrams represent the number of points per 1% of area. A, B and D are from Matanuska Glacier, Alaska; C from Skalafellsjökull, Iceland; E from Catfish Greek "Till", Ontario (from Dowdeswell et al. 1985; data from various sources).

Lambert equal area net (Phillips 1971). In this way it is possible to see whether there is an up-glacier preferred orientation (the norm) or otherwise. Some recent studies (e.g. Mills 1977, Lawson 1979, Domack et al. 1980, Dowdeswell et al. 1985, Dowdeswell & Sharp 1986) have in addition undertaken an eigenvector/eigenvalue analysis to examine statistically the strength and value of the preferred orientations for which computer programs are available. By plotting eigenvalues graphically in combination with the stereographic projections, it has proved possible to discriminate between basal melt-out till, lodgement till, sediment flows (flow tills) and waterlain tills, in order of declining fabric strength. It has even been possible to distinguish the variable effects of shearing within lodgement till, and between basal tills from highland and lowland locations.

There is little agreement on the number of clasts it is necessary to measure to obtain a valid preferred orientation. Lawson measured only 25, Mills 50, others as many as 300. The author has found that reproducible results are possible with 50 measurements, even where there is only a weak or low preferred orientation.

1.4.4 Surface features on clasts

Debris transported at the base of a glacier acquires certain distinguishing characteristics. The development of facets (flat surfaces with rounded edges) is widespread, often with as many as 80% of clasts being affected. Occasionally two parallel sets of facets give rise to "flat-iron" shapes. Other clasts may develop pentagonal or bullet shapes. Facets tend to develop on all rock types if basally transported, and they do not necessarily form preferentially along bedding, foliation or joint surfaces.

Striations and associated features, such as crescentic gouges and chattermarks, develop on basally transported stones; they are especially common on subrounded and faceted clasts. Whether or not striations develop very much depends on lithology. Hard crystalline rocks, such as quartzite, granite, gneiss and schist, rarely display striations, whereas fine-grained igneous rock, carbonates and mudstones, commonly do. For example, in glacigenic sediment from the Antarctic continental shelf Kuhn et al. (1993) found that, out of populations of several hundred, only 4% of gneisses had striations, in comparison with 43% of basic igneous rocks.

1.4.5 Surface features on sand- and silt-sized grains

Surface features on sand- and silt-sized grains are generally investigated using the scanning electron microscope (SEM). It has been demonstrated that quartz grains with sharp edges and conchoidal fracture patterns are characteristic of glacial transport (Fig. 1.6), although in lithified deposits such textures may have been altered by subsequent mechanical and chemical processes (Krinsley & Doornkamp 1973). Whether the depositional environment is marine or terrestrial has little bearing on the nature of these fracture characteristics, which may be summarized as follows (Krinsley & Funnell 1965):

23

Figure 1.6 Scanning electron microscope photographs of quartz grains from glacier ice at the margin of Grinnell Ice Cap, Baffin Island. (a) Sample from clean ice (supraglacial or englacial transport); (b) sample from basal debris-rich ice. The longest axes in both cases measure approximately 0.8 mm (photographs courtesy of J. A. Dowdeswell).

(a) large variation in the size of conchoidal breakage patterns probably caused by the variation in the size of particles ground together;

(b) very high relief (compared with grains from aeolian and littoral environments), caused by the large particle sizes present and the greater amount of energy available for grinding;

(c) semi-parallel steps, probably caused by shear stress;

(d) arc-shaped steps, probably representing percussion fractures;

(e) parallel striations of various lengths, caused by movement of sharp edges against the grains in question;

(f) prismatic patterns consisting of a series of elongated prisms that may represent cleavage and including a very fine-grained background that may indicate recrystallization;

(g) imbricated breakage blocks that look like a series of steeply dipping hogbacks;

(h) small-scale grinding indentation – irregular markings that frequently appear with conchoidal blocks.

In general, these criteria are related to processes of slow intergranular attrition under stress. However, interpretation of these textures requires care. A detailed study of a contemporary glacial environment by Whalley & Krinsley (1974) indicated that there were few diagnostic properties that could characterize any particular glacial subenvironment. The general variability of grain surfaces, from whatever position in the glacier they were taken, made it impossible to establish the source and transport mechanism of a particular grain. Freshly weathered supraglacial debris, derived by rockfall, had many of the characteristics found in subglacially transported grains. Other processes, unrelated to glacial transport, can give rise to intergranular attrition, e.g. mudflows (Harland et al. 1966). Textures of quartz grains in

24

till are also dependent on inherited characteristics, i.e. their nature in the parent rock (Whalley & Kingsley 1974). With regard to pre-Pleistocene tillites the usefulness of the technique is further limited as a result of surface-texture modification during diagenesis, and few SEM studies of such rocks have proved useful.

Microscopic studies on garnet grains have proved to be of some use in providing evidence of glacial transport. Folk (1975) first recognized trails of crescentic marks on a garnet grain in till. These trails are of parallel orientation, are uniformly spaced and of similar size. On any one surface there may be several trails differing in width, length and orientation. They can be neither the result of random impact nor attributable to an orientation effect of the internal atomic lattice. The crescentic marks on garnets resemble other glacial chattermarks (although four orders of magnitude smaller); the latter are caused by the rhythmic release of strain by fracturing as a glacier moves over bedrock. It is thus thought that chattermark trails form on garnets when they are held fast in glacier ice and slowly grind past other grains or bedrock under great stress (Folk 1975). Chattermark trails have not been observed on other mineral grains; possibly the reason is that garnets are hard, while other grains or bedrock (including quartz) are subject to crushing.

Some 15% of garnets in Pleistocene tills in North America have chattermark trails, whereas as many as 33% have been documented in Late Palaeozoic tillites from Gondwanaland (Gravenor 1979). Chattermark trails are thus useful in discriminating between glacial and non-glacial diamicts. It has been found that the statistical chance that garnets will encounter the right conditions to be chattermarked increases with the distance travelled in englacial transport, a characteristic that has further assisted in interpreting the distribution and extent of ice in pre-Pleistocene sequences (Folk 1975, Gravenor 1979, 1980). Chemical etching can also produce crescentic features that might be mistaken for chattermark trails, but if there are several crescentic features in line, chemical action is considered unlikely (Gravenor 1981).

1.4.6 Mineral composition

The mineral composition of most clastic sediments tends to reflect the degree to which they have been transported and reworked. For example, the earlier formed, high-temperature minerals in igneous and high-grade metamorphic rocks tend to be unstable at near-surface temperatures and pressures. The order of stability, starting with the most stable, is as follows: quartz, zircon, tourmaline, chert, muscovite, microcline, orthoclase, plagioclase, hornblende, biotite, pyroxene, olivine. Although chemical weathering is now recognized as an important process in the ice-free areas of the Antarctic and Arctic, and may well have been instrumental in preparing the bedrock for glacial erosion, once ice is eroding and transporting fresh bedrock, this trend of mineral stability is no longer revealed. Chemical weathering, although it does occur at the base of a glacier, is of minor significance except in areas of carbonate bedrock. However, abrasion and grinding continually produce fresh rock which, if cemented in the ice mass, is effectively preserved from subsequent chemical weathering. Supraglacial debris may be subjected to little more than frost weathering, and again the breakdown of even the most unstable minerals is slow. Till deposited after transport over long distances therefore contains a proportion of easily identifiable, fresh-looking feldspar grains, for example, whereas in other environments transport would lead to their being rapidly rounded and chemically altered.

Clay mineral assemblages in tills, obtained by X-ray diffraction techniques, are also distinctive; clay size fractions with illite and chlorite are typical of glacial environments (Alley & Slatt 1976), unless containing a high proportion of preglacially weathered material. Analysis of the clay content of glacigenic sediments allows the proportions of the major clay minerals to be estimated, which in turn allows an assessment of the contribution of weathered versus unweathered material to be made. Such data have proved to be of particular value in deep-sea sediments bordering the Antarctic ice sheet, where the glacial component may not be immediately obvious (e.g. Ehrmann & Mackensen 1992). For example, the onset of continental East Antarctic glaciation around 36 Ma is indicated by the change from smectite-dominated to illite- and chlorite-dominated assemblages, the latter indicating physical weathering under a cooler climate.

An exception to the absence of chemical alteration is where carbonate is present in the system. Work on ultra-thin sections on Precambrian tillites (Fairchild 1983) has shown that carbonate ground into a fine **rock flour** and deposited in a marine environment is prone to rapid recrystallization.

1.4.7 Geochemistry

Many analyses of whole-rock geochemistry of glacigenic sediments have been undertaken, but with rather unhelpful results – till typically has the geochemical character of a greywacke. However, one successful approach has been the application of major element chemistry to glacigenic and associated sediments in order to determine the pattern of climatic change. This approach was developed by Nesbitt & Young (1982) and applied to the diamictite-bearing Early Proterozoic Huronian Supergroup in Ontario. These authors devised a Chemical Index of Alteration (abbreviated to CIA with America's counter-espionage agency in mind!) which takes account of the fact that, during chemical weathering, feldspars degrade into clay minerals. Ca, Na and K are removed from the feldspars by soil solutions, so that the proportion of alumina and alkalis typically increases in the weathered product. The degree of weathering can therefore be quantified using molecular proportions:

$$CIA = [Al_2O_3/(Al_2O_3 + CaO^* + Na_2O + K_2O)] \times 100$$

where CaO* is the amount of CaO incorporated in the silicate fraction of the rock. This allows one to contrast glacial and non-glacial sediments in the same basin.

The geochemical character of carbonates in glacigenic sediments can also be revealing. Five distinct types of carbonate have been recognized in glacial sediments by Fairchild et al. (1989) and Fairchild & Spiro (1990): precipitation from sea or lake water, with or without microbial action; from rocks ground up by glaciers; from rapid recrystallization of this rock flour; and from groundwater near saline lakes such as those in arid regions of Antarctica. These authors further demonstrated that proportions of different isotopes of oxygen (^{18}O, ^{16}O)

provide an indication of palaeolatitude. This is based on the premise that present-day precipitation has increasingly lower proportions of the heavy isotope towards the poles.

In deep-sea sediments with or without a glacial component, in which the stratigraphic record is complete, the ratio between the heavy and light isotopes of oxygen in planktonic foraminifera may yield important climatic information that can be linked to global palaeotemperature and ice-volume changes (Shackleton & Kennet 1975, Miller et al. 1987).

1.4.8 Palaeomagnetism

The magnetic characteristics of diamicts provide useful additional information concerning depositional processes (Eyles & Menzies 1983). In lodgement tills, the magnetic particles are poorly aligned with respect to the Earth's magnetic field as a result of shear dispersion during deposition. In comparison, diamicts formed as a result of deposition in the sea or a lake show a magnetic alignment within the Earth's magnetic field with respect to azimuth and inclination. Magnetic orientation measurements are usually undertaken on small cubes or cylinders placed in a magnetometer in the laboratory.

With appropriate equipment, the magnetic fabric may be measured in the field, thus allowing ice-flow directions to be determined rapidly without the time consuming work of clast fabric analysis. Palaeomagnetism is also useful in a stratigraphic context, since particular till units may have a distinctive fingerprint. For sequences that embrace reversals in the Earth's magnetic field, palaeomagnetism may be used to provide temporal control on glacigenic sediments. However, this is of limited use in terrestrial sequences as deposition tends to be sporadic. Far better use has been made of this technique in deep-sea sequences.

1.4.9 Geotechnical properties

A variety of techniques used by engineers to assess the stability of glacigenic sediments may be mentioned briefly. These techniques allow certain geotechnical parameters to be determined and related to the geological processes responsible for them. The parameters include liquid limit, plastic limit, natural moisture content, shear strength and compressibility. The use of these parameters enables one to discriminate between, for example, different types of massive diamicts since a sheared lodgement till from which water has been squeezed will be tougher than a waterlain till. For a useful discussion of these parameters, the reader is referred to the review by Sladen & Wrigley (1983).

1.5 Evidence of glaciation in the geological record

The recognition of glacigenic sediments in the rock record is of fundamental importance to palaeoclimatology and the reconstruction of palaeo-environments. Their occurrence suggests harsher climatic conditions than the norm and, if their sedimentary characteristics are determined, it is possible to establish the presence, nature and extent of land, the direction of sediment transport and the characteristics of the depositional environment. As we have seen, till is an extremely varied material, but glacial environments are characterized by a variety of facies unique in the geological record. Environments of till deposition are varied, ranging from mountainous to lowland, lacustrine, tidal and marine. Few sedimentary criteria in themselves are sufficient to allow one to infer a glacial origin for a particular deposit; however, the association of several distinctive features can indicate not only a glacial origin but whether it is marine or terrestrial (Table 1.4).

Table 1.4 Principal criteria for establishing a glacial origin of diamict successions.

(a) Evidence for terrestrial glaciation

Abraded surfaces	striated and/or polished surfaces
	crescentic gouges; chattermarks
	striated boulder pavements
Clast-rich beds with:	irregular thickness (usually *c*. 50 m)
	lenses of sand/gravel (glaciofluvial)
	depositional shear structures in massive diamict; otherwise structureless diamict
	preferred clast orientation

Depositional fossil landforms, e.g. moraines, eskers

(b) Evidence of glaciomarine/glaciolacustrine deposition

Massive to stratified beds, often tens or hundreds of metres thick, with gradational boundaries

Dropstones in stratified units

Random clast fabric

Slight sorting or winnowing at top of beds

Association with fossils

Association with rhythmites of varve or turbidite origin

Association with resedimented deposits (debris flows)

(c) Evidence common to both environments

Variable clast lithologies

Poorly or non-sorted with wide range of clast sizes

Exotic (far-travelled) varieties of clasts

Fresh minerals

Constant mix of clasts over wide area common

Clast characteristics	shape variable from angular to rounded
	some striated and faceted surfaces
	flat-iron/bullet-nosed shapes
	calcareous crusts
	fragile clasts
	quartz grain textures; chattermarks on garnet grains

(d) Other evidence of cold climate

Ice wedge clasts

Fossil sorted stone circles, polygons and stripes

Fossil solifluction lobes

Association with lithified loess (loessite)

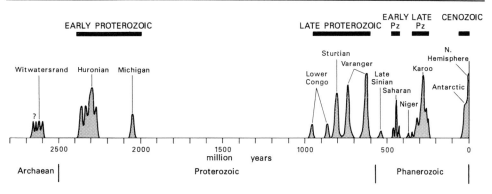

Figure 1.7   Time plot in millions of years showing periods of glaciation and their names. A qualitative measure of the extent of the Earth's surface affected is indicated by the heights of the peaks (after Hambrey 1992).

The recurrence of glacigenic sediments in the geological record suggests that ice ages are not as unusual as has sometimes been supposed (Fig 1.7, table on inside front cover). Major ice ages are represented in Early Proterozoic and Neoproterozoic, Early and Late Palaeozoic and Cenozoic strata, while other geological intervals have not been devoid of glacial activity (Frakes 1979, Frakes et al. 1992). Nevertheless, in terms of the bulk thickness of sediments, in the context of the Earth's entire history, glacigenic sediments form but a small part, although their stratigraphic value is great, especially in non-fossiliferous rock. Notable are the wide-ranging correlations made possible using tillites as stratigraphic markers for Neoproterozoic time.

Table 1.5   Typical thicknesses of drift sheets or sequences in lowland areas and continental shelves (from Flint 1971, and various other sources).

Continent	Area	Country	Thickness (m)
North America		USA	232 maximum
	Great Lakes		12
	Illinois		35
	Iowa		45–66
	Central Ohio		29 average
	New Hampshire		10 average
	Spokane Valley (Idaho/Washington)		335–442
	Fraser Delta		670
	Gulf of Alaska		5000
Europe	N. Germany	Germany	58 average
	Norrland	Sweden	4–7
	Denmark	Denmark	50
	Lubbendorf in Mecklenburg	Germany	470
	Grenoble	France	~400
	Heidelberg	Germany	397
	Imola, Po Valley	Italy	800
	East Anglia	UK	143
	Isle of Man	UK	175
	North Sea		920
Antarctica	McMurdo Sound		702 minimum
	Prydz Bay		480 minimum

Table 1.6 Distribution of glacierized areas of the world (from World Glacier Monitoring Service 1989).

Continent	Region	Area (km^2)	Totals
South America	Tierra del Fuego/Patagonia	21 200	
	Argentina north of 47.5°S	1385	
	Chile north of 46°S	743	
	Bolivia	566	
	Peru	1780	
	Ecuador	120	
	Columbia	111	
	Venezuela	3	25 908
North America	Mexico	11	
	USA (including Alaska)	75 283	
	Canada	200 806	
	Greenland	1 726 400	2 002 500
Africa			10
Europe	Iceland	11 260	
	Svalbard	36 612	
	Scandinavia (including Jan Mayen)	3174	
	Alps	2909	
	Pyrenees/Mediterranean Mountains	12	53 967
Asia	Commonwealth of Independent States	77 223	
(with all Russia)	Turkey/Iran/Afghanistan	4000	
	Pakistan/India	40 000	
	Nepal/Bhutan	7500	
	China	56 481	
	Indonesia	7	185 211
Australasia	New Zealand	860	860
Antarctica	Subantarctic islands	7000	
	Antarctic continent	13 586 310	13 593 310

Because of their widespread nature, Cenozoic and especially Quaternary glacigenic sediments have attracted by far the greatest attention. Glacigenic sediments in various guises cover 8% of the Earth's land surface, including a third of Europe and a quarter of North America (Flint 1971). These sediments have created entirely new landscapes and have provided mineral-rich soils that have influenced vegetation, land-use and the pattern of human settlement. In addition, they have provided the bulk of the sand and gravel needed by the construction industry in much of Europe, and North America.

Thus, the main focus has been on terrestrial sediments which are generally less than a few tens of metres in thickness, but in favoured locations, such as where deep basins have become filled, thicknesses may approach several hundred metres. By far the thickest accumulations of glacigenic sediment occur in marine areas, especially at the edges of continental shelves that bordered, or still border, the great ice sheets. Some examples of thicknesses of Cenozoic glacigenic sedimentary associations are given in Table 1.5.

30

1.6 Extent of glacier ice today

The distribution and extent of ice on the Earth's surface are known principally from the work of the World Glacier Monitoring Service, which has compiled a global inventory (in varying degrees of detail) of essentially all the relevant glacierized regions of the world (Table 1.6).

The large ice sheets of Antarctica (85.7%) and Greenland (10.9%) together represent 96.6% of the world's total glacierized area (13 586 310 km^2). Of the remaining 3.4% (c. 550 000 km^2), about two-thirds comprise high-latitude ice caps and ice fields, and one-third mountain glaciers. However, it is the latter which have impinged most directly on human activity, as a result of avalanches, debris-flows and outbursts of water, as well as, more positively, in providing water for hydroelectricity and irrigation (Hambrey & Alean 1992).

The principal concern today is not local glacier hazards, but whether human-induced greenhouse warming of the atmosphere will lead to melting of the polar ice sheets. The accuracy of the figures for the area of the Antarctic ice sheet (Table 1.6) is less than that for the whole of Earth's other ice masses, but the acquisition of more and better satellite data is gradually improving the picture. Current estimates suggest 80% of the world's fresh water is glacier ice, of which the greater part (30 million m^3) is in Antarctica.

If the Antarctic ice sheet were to melt totally, the sea level would rise c. 85 m (Drewry 1991). Melting of the Greenland ice sheet would add a further 7 m. The contribution of the remaining glaciers would be a mere fraction of this.

1.7 Geological timescale

Several comprehensive geological timescales have been published in recent years, all of which differ in terms of the ages of period, epoch and stage boundaries. Of most relevance to this book is the dispute concerning the onset of the Quaternary period. Internationally, the lower boundary is placed at 1.8 Ma, but most Europeans prefer one at 2.4 Ma, which corresponds to major cooling in the North Atlantic and the onset of ice-rafting. Within the Quaternary period we have the Pleistocene and Holocene epochs, the latter beginning about 10 000 years BP; these too have been much debated. In the earlier ice-age record, the Proterozoic/Cambrian boundary has been of some significance, with dates of around 570–610 Ma traditionally being assigned. Now this boundary has been placed much later at about 530 Ma, well clear of the Neoproterozoic glacial events.

For the purposes of this book, the International Union of Geological Sciences timescale of 1989 is used (see inside front cover), although maintaining some level of consistency in discussions of older glaciations has not always been possible.

2 Glacier dynamics

2.1 Introduction

Understanding some of the complexities of glacier behaviour is essential in interpreting the origin of landforms, or sequences of glacigenic sediments, on the land surface or preserved in the geological record. In the past, many authors describing Pleistocene successions would have interpreted a till horizon within a multi-layered till sequence as representing a glacier advance during a well defined cold phase; the truth is generally more complex. For example, in a glacial environment, till is intimately associated with deposits that once would have been labelled interglacial. Furthermore, glacier advances are not necessarily synchronous or related to climatic change in the short term. Indeed, some glaciers deposit most of their debris load during surges – catastrophic events related to dynamic instability of the ice – rather than to climatic change. Glacier dynamics have been reviewed in books by Paterson (1981) and Souchez & Lorrain (1991), while the highly illustrated accounts by Post & LaChapelle (1971) and Hambrey & Alean (1992) emphasize the scenic qualitites of glaciers. Only those aspects that bear on landforms and sediments are considered here. Figure 2.1 conveys a clear impression of the dynamics of the Lambert Glacier, Antarctica, supposedly the "largest glacier" in the world, as seen from space.

2.2 Formation and properties of glacier ice

Snow and ice are crystals with an hexagonal symmetry, but otherwise they occur in a wide variety of forms.

2.2.1 Derivation of glacier ice

Glacier ice is derived from: (a) recrystallization of snow during **diagenesis**, (b) melting snow and refreezing at the surface to give superimposed ice, (c) freezing of rain water, and (d) condensation and freezing of saturated air to produce rime. In temperate regions,

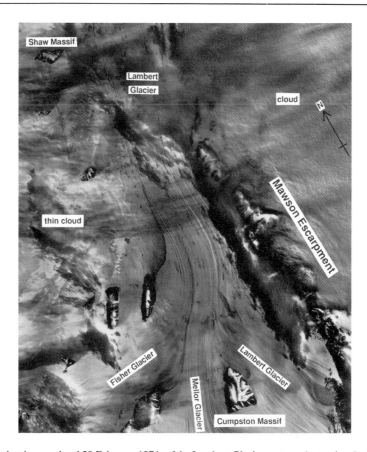

Figure 2.1 Landsat image, dated 20 February 1974, of the Lambert Glacier system, Antarctica. In the centre of the image is the strongly channelized, fast-flowing ice of three major flow units, the Fisher, Mellor and Lambert glaciers. Towards the top of the picture the Lambert Glacier passes into the floating Amery Ice Shelf, while to the left and right is the relatively slow-moving and thinner ice of the main ice sheet. The Lambert Glacier drains about a fifth of the East Antarctic ice sheet, and is the biggest valley glacier in the world, measuring 50 km across. The linear structures are medial moraines and longitudinal foliation (cf. Fig. 2.53). The distance from the top of the image to the bottom is *c.* 200 km (image courtesy of the US Geological Survey, Flagstaff, and previously published in Hambrey 1991).

recrystallization of snow at the pressure melting point is the dominant source of glacier ice, whereas in northern polar regions, where accumulation of snow is slight and summer melting occurs, **superimposed ice** (refrozen slush), is the main constituent. In the coldest and driest polar regions, notably the interior of Antarctica, no meltwater at all may be involved in diagenesis. Overall (c) and (d) are of little significance, except locally.

2.2.2 Density of glacier ice

During the diagenesis of snow, its density increases progressively. Freshly fallen snow may have a density of as little as $0.05 \, \mathrm{g\,cm^{-3}}$, snow that has survived one summer season (**firn**)

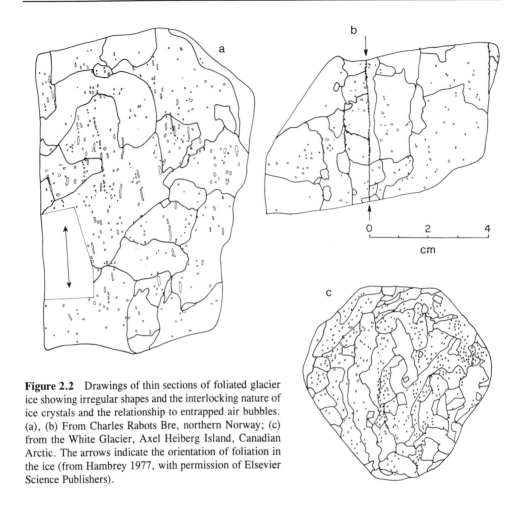

Figure 2.2 Drawings of thin sections of foliated glacier ice showing irregular shapes and the interlocking nature of ice crystals and the relationship to entrapped air bubbles. (a), (b) From Charles Rabots Bre, northern Norway; (c) from the White Glacier, Axel Heiberg Island, Canadian Arctic. The arrows indicate the orientation of foliation in the ice (from Hambrey 1977, with permission of Elsevier Science Publishers).

$0.4 \, \text{g cm}^{-3}$, true glacier ice 0.83–$0.91 \, \text{g cm}^{-3}$, and pure ice (frozen water) $0.92 \, \text{g cm}^{-3}$. As the loosely packed, randomly orientated ice crystals with intervening air spaces in firn recrystallize in response to stress, the intercrystal air spaces are eliminated, to be replaced by crystals containing bubbles of air (Fig. 2.2). The density of ice, together with the size of ice crystals, thus increases with time and depth. In the interior of the Antarctic ice sheet this is a very slow process, and it may take thousands of years, combined with burial under hundreds of metres of snow, to achieve the same effects as just a few years and a few metres of burial in temperate regions.

2.2.3 Crystallography of glacier ice

Glacier ice is analogous to deformed rocks, although it is mono-mineralic. Recrystallization is indicated by changes in the orientation of the optic- or c-axes. However, the crystal shape is often very irregular, with an interlocking character (Fig. 2.2). In glaciers of temperate regions crystals may reach 25 cm in size. The c-axis is that on which a ray of incident-parallel light is transmitted normally. c-axes are measured using a universal stage or by examining the basal planes perpendicular to the c-axis, which are picked out when bubbly ice weathers. Glacier ice often shows a preferred orientation related either to **stress** or to the **cumulative strain** axes, so it is continually modified as it moves down-glacier. Single maxima may develop perpendicular to the plane of shear when undergoing simple shear, but the pattern is often more complex, for example, with three or four maxima clustered around the normal to the shear plane. Often c-axis preferred orientations are related to deformational structures, especially the layered structure called foliation (§2.8.2). In zones of strong shear, large crystals break down into smaller crystals and generate a new foliation.

2.2.4 Deformation of a single ice crystal

The deformation of a single crystal has been studied experimentally by applying a constant stress to a single crystal orientated so that there is a component of shear stress in the basal plane, and measuring strain as a function of temperature. When stress is first applied, ice immediately deforms elastically, followed by permanent deformation (**creep**), which continues as long as the stress is maintained. Deformation is thus possible under very low stresses, and it takes place in layers parallel to the basal planes of the crystal, a phenomenon called **basal glide**. Crystals can also deform if the basal plane is not favourably orientated, but then the applied stress needs to be much higher.

2.2.5 Deformation of polycrystalline ice

Under constant stress, the strain rate initially increases with time, a characteristic known as **strain-softening**, but in polycrystalline ice many crystals are orientated so that they do not slip on the basal plane and they harden as the strain increases. If a constant stress is maintained until the total strain reaches a few per cent, the strain rate may reach a steady value proportional to a power, n (typically 1.5–4), of the stress.

If the applied stress is great enough, polycrystalline ice undergoes elastic deformation, followed by transient creep (strain rate decreasing continually, i.e. strain-hardening), then constant strain rate (secondary creep), then tertiary creep with the strain rate increasing. Overall, higher stresses are needed to induce the same effect as in a single crystal, since most crystals are not orientated for basal glide in the direction of the applied stress.

Deformation in polycrystalline ice is accomplished by:
– movement of dislocations within crystals;
– movement of crystals relative to one another;
– crystal growth;
– migration of crystal boundaries;
– recrystallization.

2.2.6 Flow law for polycrystalline ice

For glacier ice, a simple flow law has been derived which is widely used (Fig. 2.3). Many laboratory experiments have demonstrated the relationship between shear-strain rate $\dot{e}_{xy}$ and shear stress τ_{xy} over the range of stresses typical in normal glacier flow, that is up to 200 kPa (i.e. 2 × atmospheric pressure or 2 bar). A generalized flow law (known as **Glen's Law** from Glen 1952) that is applicable to glaciers relates the effective shear-strain rate e to effective shear stress in the following manner (Nye 1957):

$$\dot{e} = A\tau^n,$$

where n is a constant (discussed above) and A depends on the ice temperature, crystal size and orientation, and impurities. The flow law is now reasonably well established, but widely differing values of A and n have been obtained in different experiments (e.g. 1.5–4 for n, with a mean of 3).

It is known that disseminated debris affects the way in which ice deforms, but in a manner that is not well understood. If the volume of sand in ice exceeds 10%, the creep rate decreases with increasing sand content. Below 10% there is no clear relationship (Hooke et al. 1972). Structural evidence sometimes supports this experimentally derived observation; in a debris-rich/debris-poor layered sequence, the debris-rich layer behaves as the more rigid (**competent**) material, creating structures called **boudins** (see §2.8.2).

Ice undergoes deformation for hundreds or thousands of years. Large total strains of 100% or more are encountered. The final crystal structure may reflect this history and not just the most recent stresses.

Deformation of boreholes and tunnels, and the spreading of an ice shelf under its own weight, broadly support the flow law. Instead of using the true flow law, some authors have assumed, for simplicity, that ice behaves as a perfectly plastic material. Such material does

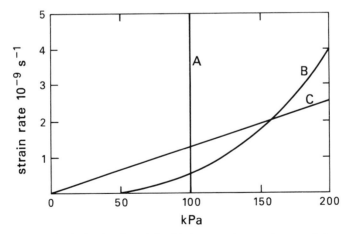

Figure 2.3 Graph showing the relationship between shear strain rate and shear stress in deformable substances. Curve A represents Newtonian viscous flow, line B perfectly plastic material with a yield stress of 100 kPa or 1 bar, curve C is for glacier ice, defining Glen's flow law (after Paterson 1981, with permission of Pergamon Press, Oxford).

Table 2.1 Classification of glaciers according to their size (given in terms of area), shape, and relationship with the surrounding and underlying topography.

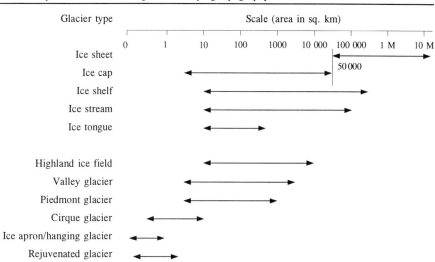

not start to deform until the stress reaches a critical value, known as the yield stress. Then the strain rate becomes very large. Stresses in glacier ice often approach 200 kPa, which approximates to perfectly plastic behaviour with a yield stress of *c*. 200 kPa.

2.3 Morphology of glaciers

Glaciers range in size from tiny ice masses only a few hundred metres across to the huge **ice sheet** of Antarctica which covers 13.6 million km^2. The global distribution of glacier ice is summarized in Table 1.6. Various names are applied to the wide range of glacier types, (Table 2.1; Hambrey & Alean 1992). At the largest end of the range is the Antarctic ice sheet, a composite of three dome-like ice sheets (Table 2.2). The bulk of the ice lies over East Antarctica, which is an elevated landmass divided by subsea-level basins bearing ice up to 4776 m thick (Fig. 2.4). The mountainous backbone of the Transantarctic Mountains separates this

Table 2.2 Dimensions of the various components of the Antarctic ice sheet (after Drewry 1983).

Region	Area (km^2)	Volume (km^3)	Average thickness (m)
East Antarctica	10 353 800	26 039 200	2565
West Antarctica	1 974 140	3 262 000	1700
Antarctic Peninsula	521 780	227 100	510
Ross Ice Shelf	536 070	229 600	430
Filchner–Ronne Ice Shelf	532 200	351 900	660
Totals	13 918 070	30 109 800	

ice sheet from the West Antarctic ice sheet, which fills a marine basin, reaching a maximum thickness of over 4000 m and a depth below sea level of 2555 m. Both the East and West Antarctic ice sheets are divisible into several drainage basins (Fig. 2.4). The much smaller ice sheet that covers the mountainous spine of the Antarctic Peninsula is dynamically separate. The combined total of Antarctic ice accounts for 91% of the world's freshwater ice and 85% of its fresh water. Huge floating slabs of ice, called **ice shelves**, buttress the East and West Antarctic ice sheets, notably the Ross and Filchner–Ronne ice shelves. There are also many other smaller ice shelves in Antarctica, but the Arctic only has small examples, in northern Ellesmere Island and Severnaya Zemlya. There appears to be a climatic control on the formation of ice shelves, the climate in the north being barely severe enough.

The only other ice sheet is that which covers most of Greenland (area: 1.7 million km²). This too is dome-shaped, filling a basin rimmed by coastal mountains to a depth of more than

(a)

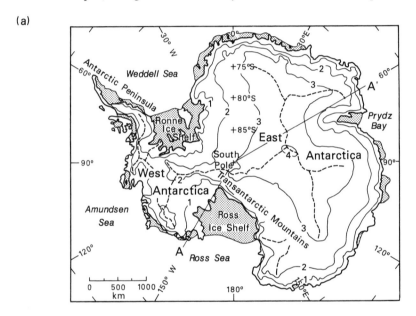

(b)

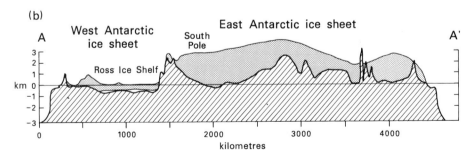

Figure 2.4 The Antarctic ice sheet: (a) ice-drainage basins (after Drewry 1983); (b) cross section drawn from ice surface and subsurface maps in Drewry (1983).

38

3000 m, from which ice flows in many places as outlet **valley glaciers** (Fig. 2.5). The Greenland ice sheet accounts for 8% of the world's freshwater ice. The ice shelves and **ice streams** of Antarctica and Greenland presently produce the majority of icebergs and are responsible for the bulk of the ice-rafted debris that enters the oceans. Whereas ice sheets are largely slow moving (typically tens of metres per year), certain parts of them attain speeds one or two orders of magnitude faster in channelized ice streams and their floating marine extensions (called ice, or glacier, tongues). The fastest ice stream is northwest Greenland's Jacobshavn Isbrae, which flows at a rate of 4.7 km yr^{-1}.

The remaining 1% of the world's ice occurs in many different topographic settings. **Ice caps** are smaller varieties of ice sheets, defined arbitrarily as covering more than 50 000 km^2 (Armstrong et al. 1973). They are characterized by relatively slow radial ice-flow. They are common in areas where the mountains have not been fully dissected and plateau remnants survive. Many examples occur in sub-Arctic and Arctic regions. **Highland ice fields** are extensive areas of undulating ice, partially reflecting the form of the underlying bed, and through which mountains project as nunataks. Outlet valley glaciers descend from these ice fields in all directions. Extensive highland ice fields occur in the St Elias Mountains on the Yukon/Alaska border, the Queen Elizabeth Islands in the Canadian Arctic, Svalbard and Patagonia.

Valley glaciers emanate from either ice sheets or highland ice fields, or are self-contained features, fed by ice accumulating on surrounding mountains (Fig. 2.6). The largest are several hundred metres thick, and they flow at rates of a few hundred metres a year. Two of the longest, the Bering and Hubbard Glaciers in southern Alaska, each flow for some 100 km from highland ice fields into the sea. Those entering the sea are **tidewater glaciers**, and there

Figure 2.5 Vestfjordgletscher, situated at the head of Scoresby Sund, East Greenland is one of the major outlet glaciers of the Greenland ice sheet seen in the background of this photograph.

Figure 2.6 The largest valley glacier in the European Alps, the Grosser Aletsch-gletscher flows 22 km from the 4000 m peaks of the Jungfrau and Mönch in the background. Note the small ice-dammed lake of Märjelensee at the bottom of the photograph.

are two types: those that are grounded on the sea bed (Fig. 2.7) and those that float (Fig. 2.8), the former being the most common. Compared with land-based valley glaciers, tidewater glaciers in fjords undergo far more pronounced fluctuations. Many subpolar and polar regions are characterized by tidewater glaciers. Some valley glaciers, especially those that originate from highland ice fields, spread out into piedmont lobes if they leave the confines of the mountains. Small **piedmont glaciers** are common in the high Arctic, but the largest, the Malaspina Glacier of subarctice southern Alaska, measures 70 km across.

Figure 2.7 The terminus of the grounded tidewater glacier, Nordenskiöldbreen, Spitsbergen. Sea ice in the foreground has been buckled into pressure ridges as a result of forward movement of the glacier in winter.

Figure 2.8 Aerial view down the heavily crevassed floating Daugaard–Jensen Gletcher, East Greenland. Note the tabular icebergs and ice-filled head of Nordvestfjord (the world's longest fjord) in the background.

On a smaller scale are **cirque** (or **corrie**) **glaciers**. They are the result of accumulation under steep cliffs on mountain flanks, eventually creating an armchair-shaped hollow which is gradually over-deepened, because of rotational ice-flow. Cirque glaciers develop preferentially on the lee side of a mountain mass where wind-blown snow accumulation is greatest. The best developed cirque glaciers occur in areas of moderate relief, especially high, dissected plateaux, as in Scandinavia and, formerly, Britain.

In contrast, high-relief areas such as alpine terrain, are characterized more by **hanging glaciers** or **ice aprons**, which are small masses of ice clinging to precipitous rock slopes. Occasionally, they shed ice avalanches of catastrophic proportions.

Lastly, **rejuvenated** or **regenerated glaciers** occur where ice-avalanche debris accumulates at a rate faster than it melts, below an ice mass that is perched on the edge of a cliff above. They are frequently conical in form, reminiscent of scree slopes below gullies.

2.4 Cold and warm glaciers

The temperature distribution (**thermal regime**) of a glacier is of fundamental importance to its dynamics, both in terms of the way the ice deforms and with regard to the rôle that meltwater plays in lubricating the bed. For example, ice *below* the pressure melting point (**cold ice**) deforms less readily than ice *at* the pressure melting point (**warm ice**), to the extent that under a given stress, ice at $0\,°C$ deforms at a rate 100 times faster than ice at $-20\,°C$.

Although transitions between cold and warm ice in the same glacier are common, it has become customary to classify glaciers thermally in the following manner.

– **Warm** (or **temperate**) **glaciers** – those in which ice is at the pressure melting point throughout except for a surface layer, *c.* 10–15 m thick, that is subject to annual cooling in winter.

– **Cold glaciers** – those in which a substantial proportion of the ice is below the pressure melting point. There are two main types: those in which all the ice is below the melting point, and so is dry and frozen to the bed, and those that undergo substantial surface melting in summer or are warmed to the melting point from beneath by **geothermal heat**. These are often referred to, respectively, as "polar" and "subpolar" glaciers, but these terms are misleading and are best avoided; even in high polar regions, glaciers undergo substantial melting, while some cold ice is present in temperate latitudes where the ice has originated from a high altitude.

The thermal regime of a glacier influences the rôle meltwater plays in the development of erosional and depositional landforms and sediments. In warm glaciers, water flows relatively freely and tends to reach the bed quickly, forming a subglacial channel and a water film that facilitates sliding. As a consequence, recycling of material by glacial streams is important. In cold glaciers with surface melting, water tends to migrate towards the margins rather than to the bed, but sliding may take place if the ice is thick enough to allow geothermal heat to raise the temperature of basal ice to the melting point. Where ice is frozen to the bed, erosion of bedrock or deposition of sediment is inhibited and no glaciofluvial sediments form. The thermal regime of glaciers entering the sea (e.g. ice shelves, ice tongues) and the resulting icebergs also strongly influences sedimentation.

2.5 Glacier hydrology

Meltwater from glaciers plays a vital rôle in the processes of erosion, the incorporation of debris into the ice and deposition. Furthermore, meltwater is a valuable resource, providing the means of generating hydroelectricity and of irrigating the land. Glaciers act as natural storage reservoirs, retaining water in winter, and releasing it in summer when it is most needed for irrigation. Meltwater emerging from a glacier can be destructive, especially when it bursts from and ice-dammed or subglacial lake, or when it mixes with loose sediment to create a debris-flow; much damage to property and loss of life has been caused in many parts of the world. In the Quaternary and older geological record, catastrophic releases of meltwater from glaciers has created impressive landscapes and shifted huge volumes of sediment. More detailed accounts of glacier hydrology are given by Paterson (1981) and Drewry (1986).

2.5.1 Sources of meltwater

Surface melting is by far the most important source of water in temperate and cold glaciers, and is supplemented by runoff from melting snow along the valley sides. Solar radiation is the principal energy source for this process. Rainfall also contributes significantly to melting on these types of ice mass. Melting rates on the lower reaches of a typical alpine glacier are $10\,\mathrm{m\,yr^{-1}}$, and $2\,\mathrm{m\,yr^{-1}}$ is more typical of cold glaciers. In many parts of Antarctica and on the Greenland ice sheet, surface melting is also prevalent, but may not be the major source of water. In these areas, as well as beneath other cold glaciers of sufficient thickness, geothermal heat is an important generator of meltwater. Water is also produced by frictional heating as ice slides over its bed, and from the effects of ice being rapidly strained, especially close to the bed.

2.5.2 Water flow through a glacier

The most noticeable aspect of water movement in the glacier system is the manner in which it flows over the surface, but water flow englacially and subglacially is equally important (Fig. 2.9). The supraglacial drainage system resembles that developed on bedrock or consolidated sediment. Dendritic stream patterns are common. Channels tend to be straight or meandering and are often incised to a depth of several metres (Fig. 2.10). Straight channels are frequently controlled by **supraglacial** moraines or ice structures such as foliation. Meandering channels may occur on exceptionally steep slopes; those on Charles Rabots Bre in Norway are found on slopes with an inclination of nearly 40°.

Supraglacial streams tend to enter an englacial position or reach the glacier bed via vertical shafts, **moulins**, which may drop sheer for several tens of metres. Frequently, moulins

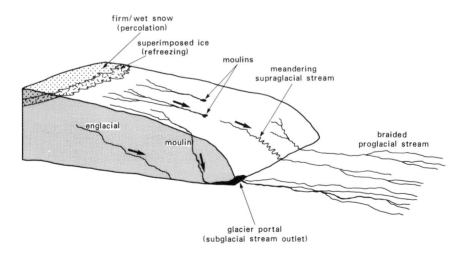

Figure 2.9 Schematic three-dimensional cross section through a temperate glacier, illustrating supraglacial, englacial and subglacial water-flow routes.

develop at sites of weakness in the ice, such as healed crevasses. On cold glaciers, supraglacial streams commonly flow towards and alongside the margins, and the resulting torrent may prove difficult to cross (Fig. 2.11). Most discharge from cold glaciers is at the margins, whereas in temperate glaciers water normally emerges from a **glacier portal** (the mouth of a cavern-

Figure 2.10 Deeply incised meandering stream flowing towards the margin of Gornergletscher, Switzerland.

Figure 2.11 Lateral meltwater stream flowing along the flank of an arm of the cold Wordie Gletcher, northern East Greenland. The stream constantly undercuts the ice margin, causing frequent collapses and temporary blockages. Note the ice debris stranded on the banks to the left.

ous tunnel at the lowest point on the bed). Otherwise, meltwater may spread out as a thin film across the bed.

Water channels at the base of a glacier are of two principal types, named after the authors who examined the physical principles of their development. **Röthlisberger channels** (Röthlisberger 1972) are pipes incised into the ice above the bed. They are convex-up in cross section, as may readily be seen from the form of the glacier portal at the **snout**, and they can be several metres high. In contrast, the smaller **Nye channels** are incised into bedrock (Nye 1973). In cross section, these features have vertical sides and a rounded or flat bottom. They tend to develop roughly parallel to ice-flow but can spiral around obstacles. Nye channels belong to the family of p-forms that are described in §3.3.2. Well defined channels do not form beneath all temperate glaciers, however. Instead, thin films form, which facilitate sliding, but which slow down the throughput of water.

A certain amount of meltwater percolates down through snow and firn, although much of it may refreeze as ice layers and ice glands, thus inhibiting further downward flow. Where firn passes down into ice, the bulk of water percolation ceases, and the firn may become saturated in summer. However, a small volume of water can also penetrate through glacier ice itself, apart from through the more obvious crevasses and moulins. Ice tends to melt preferentially at crystal boundaries, and narrow veins, a fraction of a millimetre across, develop, allowing water to pass downwards. This type of intergranular flow has been described by Nye & Frank (1973). The vein network is rarely stable, because shearing of the ice tends to sever the interconnecting veins.

During the course of a summer melt season the hydrological system undergoes considerable modification, as dye-tracing tests have shown. The combination of minimal meltwater production and ice deformation results in closure or partial closure of the previous summer's channels. Thus, the glacier is able to hold back meltwater in the early part of the melt season, resulting in high basal water pressures. As the summer progresses, the channels become more open so that, towards the end, flow is uninhibited and basal water pressures are low. In stagnant ice the channels do not close up, but progressively enlarge each summer. The end result may be a body of ice, riddled with moulins, supraglacial and englacial channels that bear a strong resemblance to the landforms of limestone solution, hence the term **glacier karst**.

Not all meltwater that reaches the bed of a glacier emerges at the snout. Some of it is forced into the underlying sediment, such as till, which thereby is easily deformed and facilitates glacier flow (§2.7.3). Some water may flow into the bedrock if it is permeable or well jointed.

During periods of rapid meltwater production, englacial and subglacial streams may be under considerable pressure. Air bubbles may implode, creating a shock that can facilitate erosion of bedrock or sediment. This process, known as cavitation, is well known from studies of dam failure where small leaks may evolve rapidly into serious breaches. Occasionally, there is a visible manifestation of water under high pressure. Outlet streams may emerge with considerable force, while at the surface fountains, or more commonly springs, may be observed.

2.5.3 Ice-constrained lakes

Lakes associated with glaciers are of four main types:
- supraglacial lakes;
- ice-dammed lakes;
- subglacial lakes;
- proglacial lakes.

Lakes are important because they commonly fill up during the summer until they become unstable and water bursts out from them as **jökulhlaups**. Ponding of water by cold ice is widespread in the Arctic, but it is also important in some temperate regions.

Supraglacial lakes are typical early melt-season features and are generally no more than a few metres deep. Many gradually empty as the drainage network opens up. In contrast, ice-dammed lakes fill slowly as the melt-season proceeds until the hydrostatic head is sufficient to allow water to flow under the ice, creating a high-discharge event (a jökulhlaup) that lasts typically for a day and leaves many icebergs stranded (Fig. 2.12). However, not all lakes burst out annually.

Figure 2.12   Ice-dammed "Between Lake", formed at the confluence of the White (left) and Thomson (right) glaciers on Axel Heiberg Island, Canadian Arctic (cf. Fig. 2.22). The photograph was taken during the jökulhlaup of 1975, the lake already having lowered several metres from the clearly defined water marks on the ice to the left and right.

Ice-dammed lakes form: (a) in the notch where two valley glaciers join, (b) where a relatively ice-free side valley enters the main glacier valley, (c) where a side glacier enters, and blocks off, drainage in the main valley (Fig. 2.13). Ice-dammed lakes, if not properly monitored, can be very hazardous; many people have died during jökulhlaups in Switzerland and South America.

Subglacial lakes are known from Iceland where high heat flow from volcanically active areas can melt large volumes of ice, as beneath the ice cap of Vatnajökull, the type-area for the generation of the jökulhlaups which devastate the coastal plains. Much larger subglacial lakes, identified in radio-echo sounding records, occur beneath the East Antarctic ice sheet, the largest approaching 8000 km^2 in area (Drewry 1986).

Proglacial lakes occur at the front of a glacier, the ice sometimes terminating as a cliff. Lakes of this type tend to grow as a glacier recedes, unless the rate of sediment filling is too great. They are generally stable unless a lower, level outlet route can be found up-glacier.

2.5.4 Water discharge from glaciers

Discharge readings have been taken below the snouts of many glaciers, especially where hydroelectricity is being generated. The pattern of discharge provides a good indication of the nature of water flow through the glacier. At any one time, discharge has two main components, which are often out of phase (Fig. 2.14a), a variable component related to weather conditions and an underlying base-flow component. Discharge has a marked **diurnal variation** (Fig. 2.14b), especially if the weather is sunny. The maximum discharge may be twice

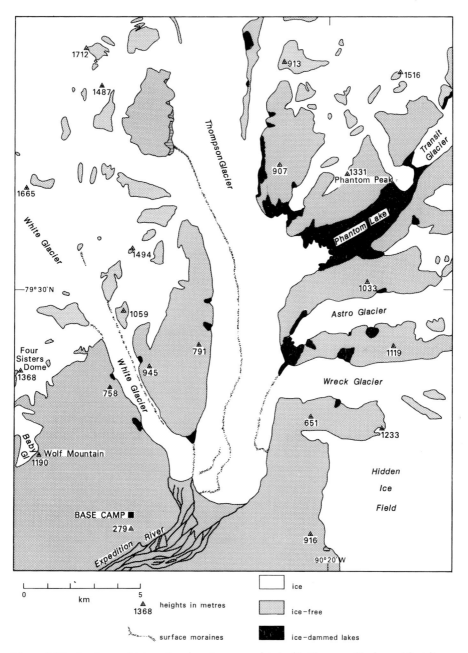

Figure 2.13 Ice-dammed lakes of various types associated with Thomson Glacier and its tributaries, Axel Heiberg Island, Canadian Arctic.

49

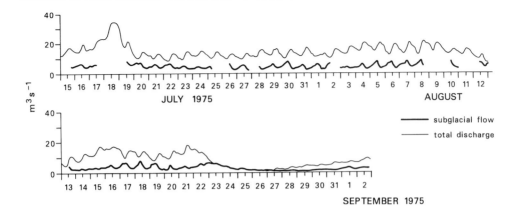

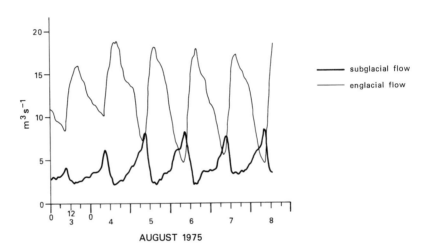

Figure 2.14 Water discharge calculated from electrical conductivity measurements in the Gornera, the stream from the Gornergletscher in Switzerland (after Collins 1977). (a) Curves showing total discharge and the proportion of discharge routed through subglacial conduits for the period 15 July to 2 September 1975. (b) Diurnal variations of the discharge components routed through the englacial and subglacial networks during sustained ablation 3–8 August 1975. Note the out-of-phase character of the peaks and troughs of the two components (reproduced with permission of Universitätsverlag Wagner Ges. M.B.H., Innsbruck).

as much as the minimum discharge or more. The marked diurnal changes have the effect of producing a flood event every 24 hours. This behaviour is an important consideration for walkers who, having crossed a gentle stream in the early morning, may have to face a dangerous torrent in the afternoon. Base-flow volumes change much more slowly. Base-flow continues throughout the winter in many cases, maintained by frictional processes or geothermal heat.

On a seasonal basis, glaciers release the bulk of a year's precipitation of snow as meltwater during the few warm months of summer. The seasonal variations normally bear little relation to precipitation; rather runoff is under a strong thermal influence. Several distinct periods of runoff on **outwash plains** below glaciers during the course of a year have been identified (Church & Gilbert 1975).

(a) Break-up of winter river ice is represented by the spring snow melt. First, saturation of the snowpack occurs, generating slush; then water begins to flow over channel ice, leading to the break-up of the winter ice on the river bed. Drainage may be well developed beneath the snow before it becomes visible.

(b) The **nival** (snowmelt) flood follows with the establishment of a connected drainage network. Much of the stored water in glaciers is released at this stage to give anomalously high flows in comparison with the daily melt.

(c) By late summer, runoff continues more or less in accordance with the degree of meltwater generation, reflecting the clearance of internal and subglacial channels in the ice.

(d) The freeze-back period of autumn follows the cessation of melting on the glacier, although draining of some channels and groundwaters is maintained. Freezing in Arctic areas may be rapid but large areas of **Aufeis** (sheets of vertically orientated ice crystals) may develop as water continues to flow from springs or from beneath the glacier into subzero temperatures. Temperate glaciers may maintain a low steady rate of flow throughout the winter.

During stages (b) and (c) above, or even at other times, the general discharge pattern may be interrupted by jökulhlaups, which produce sharp runoff peaks. Storm precipitation as rain may produce a similar effect. The discharge curve for a jökulhlaup typically shows an initial steep rise and an almost instantaneous cut-off (e.g. Whalley 1971, Theakstone 1978). High-amplitude floods are also associated with surges. This is a response to the sheet-flow of water beneath the ice, after which the glacier settles back down on its bed and the flood stops abruptly.

2.5.5 Meltwater as a geological agent

The geomorphological and sedimentological products of glacial meltwater are described in Chapters 3, 4 and 5. Here it is sufficient to mention that meltwater is both a powerful erosive agent, permitting the development of subglacial gorges, Nye channels and other p-forms (Ch. 5), and a medium for transporting and depositing large volumes of sediment. Furthermore, material carried in solution is now regarded as important in certain circumstances, especially in areas of carbonate bedrock where both rapid dissolution of rock flour and various processes of secondary precipitation of calcium carbonate can occur (Fairchild et al. 1993).

2.6 Glacier fluctuations in response to climate

Factors affecting the behaviour of the snout of a glacier are complex and numerous, but changes in climate are the most significant. The "state of health" of a glacier, or its **mass balance**, is dependent on how much snow is preserved in its accumulation area and how much ice is lost in the ablation area during the summer season. The movement of a glacier is dependent on the characteristics of the supply and loss of material. If loss exceeds supply, such as during a climatic amelioration, the whole glacier will become thinner; if the reverse is the case, the glacier will thicken in either the accumulation area (through the build up of firn) or the ablation area (through reduced melting) or both.

2.6.1 Mass balance

Changes in the mass of a glacier from year-to-year, and the characteristics of these changes spatially, are studied in order to determine the balance between gains and losses, that is the glacier's **mass balance**. Mass-balance studies are made on valley glaciers throughout the world, and on some there are continuous records going back over 30 years, for example, Storglaciären in Sweden since 1946, Hintereisferner in Austria since 1952, South Cascade Glacier in Washington State, USA since 1957, and White Glacier on Axel Heiberg Island, Canada since 1957 (Paterson 1981).

Mass balance reflects the difference between net gains (accumulation) and losses (ablation) in a given year. A positive mass balance refers to an excess of accumulation over ablation, and for a negative mass balance the reverse is true. Accumulation is represented by a variety of processes that eventually lead to the formation of ice: direct snowfall, avalanches, freezing of meltwater and slush, and the development of rime. Ablation embraces those processes which result in loss of material from the glacier: melting and the subsequent runoff of water, calving of icebergs into lakes or the sea, erosion by streams, evaporation and wind erosion.

In most cases, mountain glaciers have a clearly defined accumulation area towards their upper end, and an ablation area towards the snout. However, the influence of wind on the distribution of snow may complicate this pattern. In Antarctica the picture is rather different, since almost all ablation is by calving or through melting in contact with sea water. In some arid interior parts of Antarctica, ice streams may be subject to net ablation but, on approaching coastal areas where precipitation is greater, net accumulation may take place. Clearly, direct measurement of the mass balance of the Antarctic ice sheet is fraught with uncertainty, but as the resolution and extent of satellite data improve, continuing efforts are being made to assess the state of health of that continent's ice cover. Series of satellite images have already been used to demonstrate the rapid recession and disintegration of an ice shelf in the Antarctic Peninsula under the influence of changing oceanographic conditions (Doake & Vaughan 1990). Further work of this nature, combined with surface observations, is of vital importance if we are to understand the links between ice volume and sea-level changes.

In mass-balance terms, the surface of a glacier may be divided into a number of zones, separated by distinct lines (Table 2.3). As summarized by Paterson (1981) the zones are best developed in a cold glacier descending from high altitude, but not all glaciers have all zones. The dry snow zone is developed only on the highest mountains of Alaska and the Yukon, and in the interiors of the Greenland and Antarctic ice sheets. On some cold glaciers all the accumulation is in the form of superimposed ice, whereas on many temperate glaciers this form of ice may be negligible.

Table 2.3 Subdivision of a glacier surface into its constituent mass-balance zones.

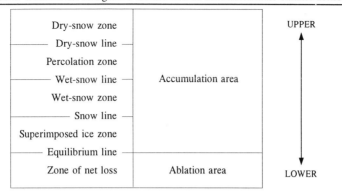

Dry-snow zone		UPPER
——— Dry-snow line ———		
Percolation zone		
——— Wet-snow line ———	Accumulation area	
Wet-snow zone		
——— Snow line ———		
Superimposed ice zone		
——— Equilibrium line ———		
Zone of net loss	Ablation area	LOWER

2.6.2 Effect of mass-balance changes on flow

Glaciers continually grow and decay in response to climatically controlled mass-balance changes. The appearance and disappearance of the great Antarctic and Northern Hemisphere ice sheets during the Cenozoic Era represent long-term changes over millions or tens of thousands of years, respectively. On the human timescale we have seen how many glaciers responded to cooling in the 18th and 19th centuries. During this period (the so-called Little Ice Age), glaciers extended far down the valleys of the Alps and Norway, for example, creating many hazards and problems for the small farming communities (Grove 1988). Since then we have seen a general recession of most temperate glaciers, in response to the subsequent warming, leaving behind a poorly vegetated zone between the ice surface and the former high ice level, called a **trimline**. Even so, glaciers react differently to climatic changes, so that in the Alps some glaciers advance at the same time as others recede; only the relative proportions vary, and a large sample of glaciers is needed to determine the overall trends. On the shortest timescale, glaciers may recede during the summer and readvance in the winter but without making up lost ground, thus eventually leaving behind a set of annual moraines.

Some of the differences in glacier behaviour may result from climatic differences between adjacent glaciers. One that begins at high altitude may benefit more from increased snow precipitation associated with higher temperatures than one at a lower altitude. Alternatively, the response times of some glaciers to changes in mass balance may be relatively long, so that small perturbations in climate during a general warming phase may not be reflected in changes at the snout, whereas other glaciers may be much more sensitive.

Changes in mass balance may be propagated down-glacier as a **kinematic wave**, which is sometimes manifested as a bulge in the ice surface. The wave moves down-glacier at a rate faster than the ice velocity, and when it reaches the snout, years later, an advance may be initiated. However, the distance a glacier will advance is determined solely by the amount of available ice, not by the kinematic wave itself (Paterson 1981).

In many cases, it is the glacier snout which is extremely sensitive to mass-balance changes, whereas the middle and upper reaches of the glacier may show little variation in ice thick-

ness. The position of a glacier snout may be a good indicator of climatic changes because of this sensitive behaviour. However, in some cases, the snout does not recede, but the entire tongue may waste downwards as the ice stagnates. Such circumstances prevail when the supply of ice from above is reduced drastically over a short period.

The response time to mass-balance changes varies dramatically from place to place. On the one hand, the glaciers that flow towards the west coast of South Island, New Zealand, respond within a few years to the marked changes in accumulation that occur in this region of huge precipitation. Typical glaciers in the European Alps and the Western Cordillera of North America may respond within a few decades. In contrast, the Antarctic ice sheet with its low precipitation may take many thousands of years to respond to changes in mass balance. The latter assessment makes it very difficult to determine how the large ice masses may change in response to human-induced global warming.

2.7 Glacier flow

Glaciers flow in two main modes, commonly in combination: **internal deformation** and **basal sliding**. The combined effect of these two methods on the displacement of markers on the surface and within the ice is depicted in Figure 2.15. A third mechanism – movement over a bed of deforming sediment – is also important in certain circumstances. The cumulative result is a velocity distribution pattern showing in general terms:

(a) an increase in forward velocity towards the equilibrium line and decrease below;

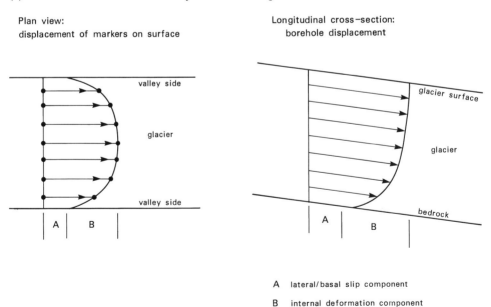

Figure 2.15   The two main components of ice-flow in a valley glacier: lateral/basal sliding and internal deformation.

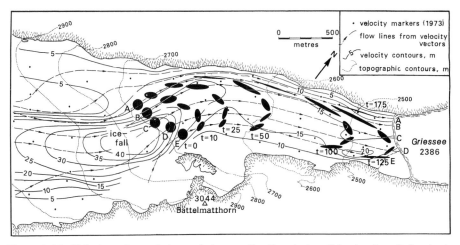

Figure 2.16 Velocity and cumulative strain in a small valley glacier, Griesgletscher, Switzerland (Hambrey et al. 1980). Note the increase in velocity (to $>40\,\mathrm{m\,yr^{-1}}$) and convergence of flowlines towards the midway point (coinciding with the equilibrium-line and the top of an icefall), and an irregular decrease with near-parallel flowlines below. The cumulative strain pattern is indicated by the progressive development and rotation of ellipses derived from circles drawn at arbitrary points. Note that the marginal ice is subject to much greater cumulative strain than ice in the centre (reproduced with permission of the International Glaciological Society).

(b) an increase in velocity towards the centre, with the strongest velocity gradient near the margins;
(c) a decrease in velocity towards the bed with the strongest velocity gradient near the base of the glacier (Fig. 2.15);
(d) convergence of particle paths (flowlines) in the accumulation area and divergence in the ablation area.

Figure 2.16 illustrates a typical velocity distribution pattern in a small Alpine valley glacier.

2.7.1 Internal deformation

We have considered above how ice deforms at the crystal scale and the influence this has upon the way in which a glacier flows. We now look at how ice deforms on the scale of the entire glacier. Flow broadly is the result of gravity forces which induce spreading of the ice by creep under its own weight (Drewry 1986). In the simplest case of a parallel-sided slab of ice which does not slide, and in which movement consists only of deformation within the slab due to iis weight, deformation is said to be laminar, and flowlines are parallel to the top and bottom of the slab. More often, however, the top and bottom surfaces of the glacier are not parallel, and the ice flows in the direction of the maximum surface slope, even though the bed slopes in the opposite direction. This explains such phenomena as over-deepened basins in which basal ice must have flowed uphill. Bedrock irregularities disrupt the laminar flow pattern and, as a result, bed-parallel structures may become folded.

Further complications are introduced by longitudinal variations in velocity. In this situation, two flow states are possible: **extending flow**, in which the upper part of the glacier is tensile as a result of the ice accelerating; and **compressing flow**, in which the ice is compressive at all depths as a result of deceleration (Nye 1957). These two types of flow give rise to distinctive crevasse patterns in valley glaciers (Fig. 2.17) where there is a strong velocity gradient at the margin as well as longitudinally. The thickness of the ice changes in response to extending flow (thinning) and compressive flow (thickening).

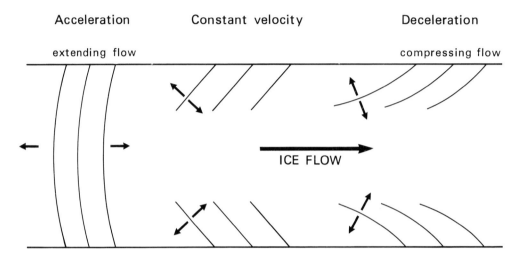

Figure 2.17   Flow states in a valley glacier (after Nye 1957). Extending flow parallel to the direction of ice movement occurs where ice accelerates, compressing flow where it decelerates. The small arrows indicate the orientation of the maximum tensile stresses, and are normal to the crevasses marked.

Extending flow occurs in a valley glacier in the accumulation area as mass increases, where the bed steepens (an icefall being the extreme case), where it is joined by a side stream in a confined channel, and on the outside of a bend. Compressive flow occurs where the bed becomes gentler, where the valley widens, on the inside of a bend, and towards the snout as mass is lost.

The above considerations apply to the deformation that is normally measured on a short timescale (months to years). However, the cumulative effect of deformation over a long period (tens, hundreds or thousands of years) also needs to be examined, as it is the total amount of deformation (referred to as **total** or **finite strain**) that explains the disposition of many structures and debris within the ice.

Cumulative strain can best be visualized using the strain ellipse, as is common in the analysis of deformed rocks. Beginning with an arbitrary circle, deformation along a flowline or particle path changes it into an ellipse. Using a velocity distribution and flowline map, and assuming steady-state flow, Hambrey & Milnes (1977) derived the cumulative strain pattern of the ablation area of Griesgletscher in Switzerland (Fig. 2.16). In the centre of the glacier, ellipses formed initially with long axes perpendicular to the flowline (**pure shear** or **non-rotational strain**) below an icefall, then showed little change thereafter. In contrast, near the margins, ellipses developed at a moderate angle to the flowlines, but gradually approached parallelism as the ice moved downwards (**simple shear** or **rotational strain**). Cumulative deformation near the snout was much greater at the margins (ellipse axial ratios typically 70:1 after 175 years, compared with that in the centre (3 to 5:1) after 125 years. Hudleston & Hooke (1978) investigated the same phenomena in a depth section (Fig. 2.18). Modelling deformation along a 10 km flowline in the Barnes Ice Cap, Baffin Island, the strain ellipses showed compression normal to the particle path in the upper part of the ice cap (pure shear, or non-rotational strain), but gradual rotation of the ellipse towards parallelism with the particle path with increasing depth.

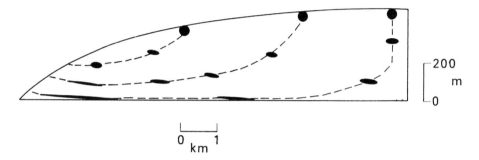

Figure 2.18 Deformation of circles into ellipses along flow lines through the Barnes Ice Cap, Baffin Island, shown in a vertical longitudinal cross section. These ellipses were calculated using finite-element computer modelling (from Hooke & Hudleston 1978, with permission of the International Glaciological Society).

2.7.2 Basal sliding

Whereas internal deformation affects all glaciers, not all slide on their beds. Nevertheless, all warm glaciers and many cold glaciers do slide, the latter in response to geothermal heat flux from bedrock if the ice is sufficiently thick to act as an insulating blanket from the ambient air temperature. Extensive areas of basal ice in the Antarctic and Greenland ice sheets are at the pressure melting point and slide. Most fast-flowing valley glaciers and ice streams attain their high velocities because basal sliding is the major component of the total velocities. The basal-sliding component has been recorded for several glaciers, examples being 50% for the Grosser Aletschgletscher in Switzerland, 90% for the Blue Glacier in Washington State (both valley glaciers), 9% for Vesl Skautbreen, a cirque glacier in Norway, and 0% for the Meserve, a glacier frozen to its bed in Antarctica. The highest percentages are associated with rapid melting or heavy rainfall, when the bed is lubricated. Basal sliding is largely responsible for the bulk of erosion, transport and deposition of debris taking place.

   Basal processes have been subject to several rigorous theoretical analyses, notably by the physicists Weertman, Lliboutry and Nye (Paterson 1981: Ch. 7), but the precise mechanisms are still far from fully understood. Three main processes have been recognized at the bed of the glacier. First, enhanced basal creep occurs in the lowest few metres of the glacier, a process possible in ice at any temperature, but one that is more effective in warm ice than in cold ice. This process allows ice to flow around an obstacle, such as a boulder on the bed, and determines the direction in which basal debris is transported. A second mechanism, involving

pressure melting and **regelation**, occurs at the ice/rock interface. When ice moves over a bump in the bed, pressure melting occurs on its upstream side, the water flows over and around the bump to the lower pressure zone, where it refreezes and forms regelation ice, possibly in a cavity, and at the same time loose debris becomes attached to the **glacier sole** (Fig. 2.19). The critical maximum size of the bump is debatable, but a figure of 1 m has been postulated. This process has been observed under many active glaciers. A third mechanism involves slip over a water layer. Water at the base of a glacier not only "lubricates" the bed but also smooths out the smaller irregularities. A layer of water only a few millimetres thick could increase the sliding velocity by up to 100%. Much of this water may originate from supraglacial melt-water streams reaching the bed via moulins and crevasses. Direct observations in subglacial cavities beneath temperate glaciers confirm the validity of these mechanisms, but in addition suggest that the motion is jerky (e.g. Theakstone 1967, Vivian & Bocquet 1973).

From the above it will be apparent that bed roughness and the rôle of debris have an important influence on the sliding velocity. If the bed is too rough, because of bedrock irregularities, debris between ice and rock, and debris within the basal ice, the basal-sliding velocity is appreciably reduced. The difficult problem of developing a theory that takes these factors into account has yet to be solved.

Figure 2.19 Regelation of ice at the sole of the Glacier de Tsanfleuron, Switzerland, observed in a cave near the snout. Successive regelation layers with basal debris have built up in this vertical section which is about 20 cm high.

2.7.3 Deformable beds

The "effective bed" beneath a moving ice mass may not necessarily be the ice/bedrock or ice/sediment interface, since the material beneath the ice may also be deforming. In recent years it has become widely recognized that large areas beneath glaciers and ice sheets are underlain by unconsolidated sediment which fundamentally affects the dynamic behaviour of the ice mass. In essence, the deforming sediment forms part of the flowing ice mass, and the "effective" bed is the surface below which there is no forward motion.

If the base of the ice is at the pressure melting point, the underlying sediment may be satu-rated with water, and thus be prone to deformation, especially if pore-water pressures are high. Boulton (1979) examined deformation in till beneath Breidamerkurjökull in Iceland, and found that this process accounted for 90% of the forward movement of the glacier, even though the deforming layer was only about 0.5 m thick. Displacement was about 0.5 m after 10 days. The strongest deformation appeared to be in the upper part of the unit (Fig. 2.20).

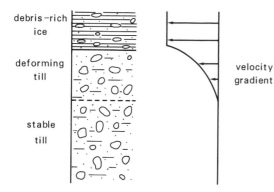

Figure 2.20 Velocity vectors parallel to flow and associated displacements in the upper part of a deformable substrate of till below the Breidamerkerjökull, Iceland. Shape of profile generalized. Displacement about 0.5 m after 10 days (after Boulton 1979).

A similar process has been inferred for one of the Antarctic's major ice streams, named "B" (Alley et al. 1989). It is conceivable that for large parts of ice sheets and valley glaciers, especially where flowing fast, deformation in subglacial sediment is the dominant component of flow. A direct result of this process is that substantial volumes of till may be transported beneath the ice by shear within the deformable layer. In the case of Ice Stream B, which merges with the Ross Ice Shelf and other ice streams, the end result of the process is inferred to be deposition at the grounding-line, which in effect represents the end of the debris "conveyor belt", giving rise to **diamict aprons** (a feature described from elsewhere in Antarctica by Hambrey et al. 1991) or in a genetic sense **till deltas** (Alley et al. 1989).

The bed may also deform if the sediment is frozen to the ice, as when a glacier overrides permafrost. In such cases blocks or large **rafts** of frozen sediment may be incorporated into the base of the ice mass. This frozen sediment may deform in a manner that resembles the flow of ice (the flow laws are similar), becoming folded and fractured in the process.

2.7.4 Long- and short-term variations of glacier flow

Glaciers show major changes in surface velocity through time. On a timescale of several years or more, velocities change in response to climatic factors. Velocities increase in response to gains in mass, following several years of positive mass balance. By the same token, a succession of negative mass-balance years results in declining velocities and, if severe, to eventual stagnation of the ice. Exceptions to such behaviour occur in the special case of surge-type glaciers (§2.9), or when a subglacial volcanic eruption takes place.

Velocity variations on a seasonal or daily timespan are largely related to meltwater production and subglacial water pressures which control the rate of basal sliding. Many studies have been undertaken on valley glaciers (see Paterson 1981: Ch. 7). Typically, during the winter, when meltwater production largely ceases, ice-flow tends to close drainage channels. During early summer the first meltwater is forced into small cavities and water pressure becomes high. Thus, ice-flow tends to reach a peak at this time, even though meltwater production does not reach a peak until a month or so later, but by this time the cavities and channels have opened up and water pressure is low. Similarly, sliding velocities reach a minimum a few months after the time that runoff ceases and water pressures are lowest. As an exam-

Table 2.4 Classification of structures in glacier ice.

Category	Structure		Type of deformation
Primary structures	Sedimentary stratification Unconformities Ice layers, lenses, glands Regelation layering (basal foliation) Ice breccia		No deformation
Secondary structures	Folds Foliation Boudinage Crevasse traces (tensional veins) Shear zones		"Plastic" deformation
	Ogives		"Brittle" deformation
	Closed fractures	normal faults strike-slip faults thrusts crevasse traces	
	Open fractures	transverse crevasses marginal crevasses longitudinal crevasses splaying crevasses *en echelon* crevasses basal crevasses *Bergschrund* *Randkluft*	

60

ple, measurements over a two year period of the temperate valley glacier, the South Cascade, showed the lowest sliding velocity in November (120 mm day^{-1}) and the highest velocity in June (220 mm day^{-1}); an 83% difference. The creep velocity (comprising internal deformation) in this case, by contrast, varied only from 40 to 50 mm day^{-1}; a 25% difference. Minimum and peak discharges occurred in March/April and July, respectively.

Velocities also vary diurnally. The bulk of surface melting is due to solar radiation, so the supply of meltwater to the bed and associated basal water pressures and sliding velocities are highly dependent on the weather. In clear weather, velocities may vary by as much as 100%, with peaks in mid-afternoon and minima in the early morning. On cloudy or rainy days the diurnal fluctuations are largely suppressed.

2.8 Structure of glaciers and ice sheets

Glacier ice is like any other type of geological material in comprising strata that progressively deform to produce a wide range of structures. The end product is a metamorphic rock that in temperate glaciers has deformed close to the melting point, and the original structures may be totally obliterated.

Two main categories of structure occur in glaciers (Table 2.4):
 - primary structures, which result from deposition or accretion of new material;
 - secondary structures, which result from deformation.

2.8.1 Primary structures

The dominant structure in the first category is **sedimentary stratification**, which is the result of accumulation of snow year by year (Fig. 2.21a). Most snow turns into firn, then coarse, bubbly ice, but summer surfaces are often indicated by a refrozen melt-layer of bluish ice and dirt. Regularly layered ice is the product, although periods of excessive ablation, which remove several or many previous layers, give rise to **unconformities**. Within the snowpack, downward percolation of meltwater takes place during the summer. Refreezing of the meltwater creates ice layers, and ice lenses or pipe-like structures called ice glands.

Another type of primary structure, **regelation layering**, is the result of pressure melting and regelation processes at the base of a glacier. It is often called basal foliation, but is not strictly a deformational structure. Commonly, it is parallel-laminated and has variable amounts of basal debris associated with it (Fig. 2.19).

Where the glacier margin is subject to collapse, especially in the neighbourhood of ice cliffs and crevasses, ice debris may reconsolidate to give a **breccia**. This structure may be welded together by trapped snow or by the refreezing of meltwater.

Figure 2.21 Structures in glacier ice. (a) Sedimentary stratification, deformed by flow and exposed on a 30°–40° slope on Charles Rabots Bre, Okstindan, Norway. Each layer represents a year's accumulation of snow above the firn line. (b) Isoclinal z-fold associated with thrusts (the dirty layers) in the terminal cliff of Thomson Glacier, Axel Heiberg Island, Canadian Arctic. The wavelength of the fold is estimated to be about 5 m. (c) Typical well developed longitudinal foliation in the White Glacier, Axel Heiberg Island, Canadian Arctic; note the supraglacial meltstream flowing parallel to the ice.

2.8.2 Secondary structures resulting from ductile deformation

Secondary structures include a range of features that can be classified broadly as the products of plastic flow and brittle fracture. Folds on all scales are commonly observed on the glacier surface and they involve both primary and secondary layered structures (Fig. 2.21b). Huge folds, many kilometres across, are well known from the piedmont glaciers of Alaska, notably the Malaspina and Bering (Post & LaChapelle 1971), while small folds, only a few centimetres across, can be distinguished in foliated ice. Normally, these folds fall into the **isoclinal** or **similar** categories, but occasionally **parallel folds** occur, reflecting more "brittle" folding (Hambrey 1977). Refolded folds are quite common. **Recumbent folds** are often present in basal ice, and they reach several metres in amplitude in cold glaciers. Hudleston (1976) described fine examples from the Barnes Ice Cap.

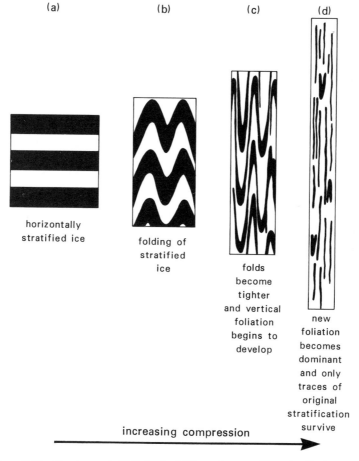

Figure 2.22 Development of foliation by folding and transposition of earlier layers (stratification or another foliation). Note how the increasing compression from (a) to (b) is accompanied by a marked change of overall shape. The foliation develops out of those parts of the original layers that are thinned the most; its final orientation is totally different from that of the original layers.

Foliation is a layered structure, comprising coarse-bubbly, coarse clear (blue) and fine (white) ice (Allen et al. 1960), that forms during flow (Fig. 2.21c). Coarse ice has crystals in the range of 1–15 cm, whereas fine ice crystals are usually less than 5 mm in diameter. Foliation is also defined by elongated air bubbles, discrete planes of shear and flattened ice crystals. Foliation generally develops from pre-existing layers, notably stratification and the traces of former crevasses (Hambrey 1975, Hooke & Hudleston 1978). Folding of the layers and progressive attenuation of the folds allows the new structure to develop on the limbs of the sheared folds. Ultimately, only isolated fold hinges may remain. This process is known as transposition and is depicted in Figure 2.22; it is also a characteristic feature of metamorphic rocks, such as schists and gneisses. Rarely, foliation appears to be a totally new structure, forming a sort of cleavage that has an **axial–planar relationship** to similar folds. Foliation develops essentially under two basic deformation regimes – pure shear and simple shear – which affect the pre-existing structural inhomogeneities in different ways (Fig. 2.23).

Pure shear is represented by normal compression and equal expansion in the direction of flow (Fig. 2.23, rows A to B). So the structures become simply flattened, a process typical of the upper and middle parts of a glacier. Simple shear results from the strong velocity gradient between bedrock and the ice, so is most prominent near the margins and the bed. In essence, the structures gradually rotate towards parallelism with the direction of flow (Fig. 2.23, rows A to C). The same effect can be demonstrated by drawing a circle or other shape on a stack of cards, and smearing them out gradually. Foliation is therefore the product of cumulative strain (Hambrey & Milnes 1977, Hooke & Hudleston 1978).

Most foliation falls into one of two main geometrically arranged categories: longitudinal and transverse. Longitudinal foliation is normally steeply dipping and it occurs at the margins of a glacier, or in association with medial moraines. In some glaciers, where ice from a wide accumulation basin feeds into a narrow tongue, longitudinal foliation may extend across the entire width and be predominantly parallel to flow (Hambrey & Müller 1978). Isoclinal

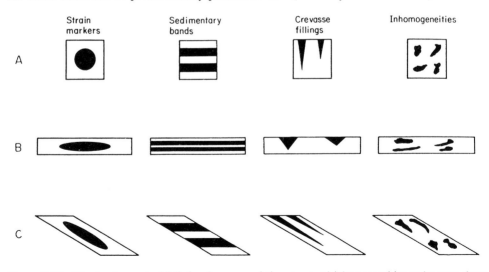

Figure 2.23 The development of foliation from pre-existing structural inhomogeneities under pure shear and simple shear regimes (from Hooke & Hudleston 1978, with permission of the International Glaciological Society).

(a)   (b)

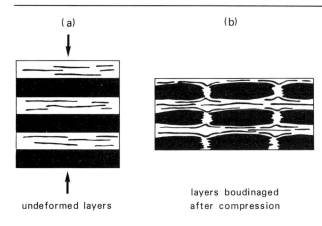

undeformed layers   layers boudinaged
after compression

Figure 2.24 Schematic diagram showing the development of boudinage structures in a sequence containing layers of contrasting ductility or competence. The boudins form in the more competent layers, the less competent material on either side flowing into the spaces between the boudins (after Ramsay 1967). In glaciers, different ice types, for example, fine-grained and coarse-grained, have different ductilities.

folds may have axes parallel to the foliation, and the entire structure may be the result of isoclinal folding of primary stratification and eventual transposition. Arcuate foliation originates in regions of transverse crevasses or **icefalls**, forming first as **crevasse traces** (described below), before being compressed in a longitudinal direction below the icefall and deformed by flow into an arcuate pattern with the apex down stream. The geometry of arcuate foliation resembles that of a set of nested spoons (Allen et al. 1960). As the foliation moves down-glacier, its up-glacier dip declines from near-vertical to a few tens of degrees near the snout. In both longitudinal and transverse foliation, the total strain ellipse is parallel to the structure, but in the former case this has been achieved mainly through simple shear, and in the latter by pure shear.

Boudins are sausage-shaped features that result from compression of a multi-layered sequence with layers of contrasting competence (or ductility) (Fig. 2.24). They are widespread in glacier ice (Hambrey & Milnes 1975) and two main types may be distinguished: competence-contrast boudinage and foliation boudinage. The former type (Fig. 2.24), is evident in ice with distinct layers of coarse and fine ice, or in dirty and clean ice. Fine ice and dirty ice often appear to behave as the more competent (less ductile) material; however, the latter observation conflicts with the behaviour of debris-bearing ice observed experimentally in the laboratory. Foliation boudinage is the more common type, being intimately associated with longitudinal foliation. Commonly, the foliation has an asymmetric appearance, suggesting formation in a rotational strain regime.

2.8.3 Secondary structures resulting from brittle deformation

A wide variety of fractures result from the brittle failure of ice, mainly in the upper 30–50 m of a glacier. Foremost among them are **crevasses**, which are usually classified according to their orientation with respect to the ice-flow direction: transverse, marginal, longitudinal, splaying and *en echelon* (Meier 1960) (Figs 2.17 & 25). In addition there are the little-studied basal crevasses and **bergschrunds**, which are the crevasses that separate the flowing ice at the head of a valley from the ice that adheres to the rock above. Crevasses may range from a few metres in length, to hundreds of metres in a large valley glacier. Some crevasses in

Figure 2.25 Accumulation-area crevasses of the mainly transverse and en echelon types in the complex upper basin of the Fox Glacier, New Zealand; bergschrunds can be seen in the uppermost reaches of the glacier.

Figure 2.26 The Hochstetter Icefall of Mount Cook, feeding the Tasman Glacier of New Zealand. The photo was taken prior to a major rockfall that obliterated much of the ice in this view (cf. Fig. 2.33).

Antarctica are over 100 km long. Crevasses vary in width from a few millimetres to several tens of metres. Some of those in Antarctica are large enough to swallow large over-snow vehicles.

Crevasses form only where at least one of the principal stresses is tensile, and where this stress is greater than the tensile strength of the ice. Crevasses normally open in the direction of this stress (Meier 1960), but, if the ice is passing through a changing stress regime or contains structural inhomogeneities, this may not be strictly true. Similarly, the stresses needed to initiate a fracture may not always be the same.

The depth of crevasses is dependent on the plastic flow of ice at the bottom. In temperate glaciers, crevasse depths in excess of 30 m have rarely been measured, an observation that is compatible with theoretically calculated depths, yet mountain literature abounds with examples of crevasses "hundreds of feet deep". Cold glaciers, on the other hand, have crevasses that are considerably deeper. Crevasses may also attain a greater depth if they are water-filled (Robin 1974), in which case they can penetrate to the bed of the glacier.

Icefalls are steep zones in a glacier where the entire surface is broken up by crevasses of many orientations. Such zones are difficult or impossible to traverse safely on foot (Fig. 2.26).

Crevasse traces is the term used for a wide family of initially vertical layers that form parallel to, or extend from, open crevasses (Hambrey 1975). One type is the product of brittle fracture, without separation of the two walls. Away from the fracture plane, ice recrystallizes adjacent to the plane under continued extension (Hambrey & Müller 1978) creating a feature analogous to tensional veins in rocks. A second type of crevasse trace is the result of freezing of meltwater in a crevasse and its subsequent closure. These usually occur as prominent blue ice layers, whereas those of the first type are more subtle. Crevasse traces survive ablation remarkably well, cropping out all the way to the snout in many valley glaciers, suggesting that deep-seated fracture of the bed may occur in both cold and temperate glaciers.

Clean-cut but unopen fractures may sometimes be observed in glaciers. **Normal faults**, with the downglacier wall displaced downwards relative to the surface by a few metres, form in ice-falls and other crevassed regions, and are visible in the walls of crevasses or along the sides of the glacier. **Strike-slip faults**, showing lateral displacements of up to a metre, are commonly seen at the surface of the glacier and may be associated with bending of the ice, resulting from shear prior to fracture. In both cases distinctive features need to be displaced, in order to recognize these movements.

Thrusts, alternatively referred to as thrust faults or **shear planes**, have been observed along the margins and near the snouts of many glaciers. They are associated with strong compression in the ice, such as where the glacier impinges on an obstacle or where the ice is slowing down. Although small thrusts may be observed in many temperate glaciers, they are best developed in cold ones. Surge-type glaciers also have many thrusts. The high Arctic is particularly well endowed with thrusts and their development is facilitated by:

(a) the rotation by flow of transverse crevasse traces into an attitude (up-glacier dip of c. 45°) that promotes displacement along the pre-existing planes of weakness;

(b) the transition from a basal-sliding regime of thick ice to one in which the ice is frozen to the bed, such as near the snout where the rôle of geothermal heat is lessened under thinner ice.

Thrusts play an important rôle in the recycling of basal debris (as discussed in §2.10.2) and they tend to be associated with recumbent folds.

2.8.4 Secondary structures of composite origin

A final type of secondary structure is referred to as **ogives**, which are alternating light and dark bands, or waves, extending in arcuate fashion across the glacier surface below some icefalls (Fig. 1.1). Each pair of light and dark bands represents a year's movement through the icefall, and where they extend to the snout they are useful measures of the transit time of ice through the glacial system. Various hypotheses have been proposed for ogive development, but the most likely is that of Nye (1958), who suggested that the ice is thinner in the icefall in summer because of ablation, and collects more dust, thereby creating the darker layers and troughs. In contrast, in winter, snow fills the crevasses, the ice thickens, so creating the light bands and wave crests. Close examination of ogives in Bas Glacier d'Arolla, Switzerland, by the author, indicates that they are composed of typical arcuate foliation, suggesting that these structures are genetically related. The darker layers thus represent more intensive crevasse and crevasse trace formation in summer.

2.8.5 Structural patterns in glaciers and ice sheets

Most structural glaciological work has been undertaken on cirque or valley glaciers, and several detailed maps, combining air-photograph information with ground measurements, have been published. In some cases, we can see how stratification evolves into longitudinal foliation, in others how arcuate foliation develops out of transverse crevasse fields. Foliation is generally strongest near the margins of a glacier and where associated with a medial moraine. In cirque glaciers, stratification may be the dominant structure throughout, but in many valley glaciers this structure is soon obliterated.

Satellite imagery provides a means of deciphering the surface structure of the Antarctic and Greenland ice sheets. Much Antarctic ice is discharged into the Southern Ocean via ice streams and ice shelves, and often shows supposedly flow-parallel features, which have been described as flowlines (e.g. Crabtree & Doake 1980, Swithinbank 1988). By comparing these Landsat images with air photos and ground observations of exposed ice, we have been able to interpret these linear structures as longitudinal foliation resulting from the probable passage of stratified ice into confined channels, thereby subjecting it to isoclinal folding (Reynolds & Hambrey 1988, Hambrey & Dowdeswell 1994). Much can be inferred from Landsat-derived structural maps about flow dynamics and sediment-transfer paths. Figure 2.1 shows longitudinal foliation in one of the largest ice-drainage systems in Antarctica, the Lambert Glacier system.

In conclusion, although glacier ice is rheologically simpler than most rocks, being mono-mineralic, and deforming solely under the influence of one driving force, gravity, the structural pattern may be exceedingly complex. However, because ice deforms at a rate that can be measured and allows estimates of total strain to be made, glaciers can serve as models for deformation deep within the Earth's crust. For example, the same range of structures occur in glaciers as in mountain belts, such as the Alps, even though the former deform six orders of magnitude faster (Hambrey & Milnes 1977).

Structures in glacier ice tell us much about the nature of deformation on both short and long timescales. They can also be used to infer flow paths of debris, and explain the disposition of debris on, within, and below, the ice.

2.9 Glacier surges

Glacier surges are a form of exceptionally rapid flow that occurs for a short period (typically a few months) in certain glaciers, and is preceded and followed by longer intervals (typically many years) of quiescence when the ice is relatively stagnant. According to Sharp (1988a), about 4% of glaciers are of the surge type, but their distribution is restricted to certain areas, suggesting that their environmental setting may have some influence (Raymond 1987, G. K. C. Clarke 1991). Both temperate and cold valley glaciers may surge, and it has even been suggested that large parts of the Antarctic ice sheet may also surge, a scenario which, if true, could have globally catastrophic consequences in terms of sea-level and climatic changes. Surge-type glaciers have two distinct zones: the reservoir and the ice-receiving area, which do not coincide necessarily with the normal accumulation and ablation areas.

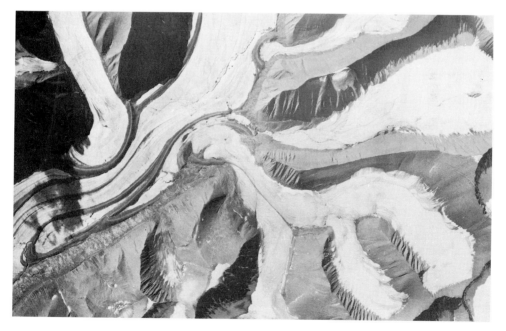

Figure 2.27 Surge-type glacier complex in Spitsbergen (Battybreen) during a period of quiescence (1969). Irregular pulses of ice-flow give rise to their distinctive looped and teardrop-shaped medial moraines (vertical air photograph no. RC8 S69 1466, courtesy of Norsk Polarinstitutt).

69

2.9.1 Characteristics of surge-type glaciers

Glacier surges and surge-type glaciers have been studied in detail in several locations, including Alaska, the Yukon, the Pamirs and Svalbard. The processes involved during a surge event are therefore quite well known, but what actually triggers the surge, and the nature of the bed over which the ice flows are the subjects of much debate. Excellent reviews of glacier surges have been provided by Kamb et al. (1985), Raymond (1987) and Sharp (1988a,b).

When a glacier surges, it displays several features:

(a) its surface becomes very heavily crevassed;

(b) ice velocities increase 10- or a 100-fold (to $5\,km\,yr^{-1}$, for example) and fluctuate wildly;

(c) large volumes of ice are transferred from the upper receiving basin to the lower part of the glacier, resulting in a net reduction of the surface gradient and leaving ice stranded at the higher levels;

(d) the snout often (but not always) responds with a rapid advance, typically a few kilometres over a few months;

Figure 2.28 Surge-induced thrusts and folds in the tongue of Variegated Glacier, southern Alaska. The photograph was taken in 1986, three years after the glacier's most recent surge. The roughness of the ice surface is a relict of the intense crevassing induced by the surge.

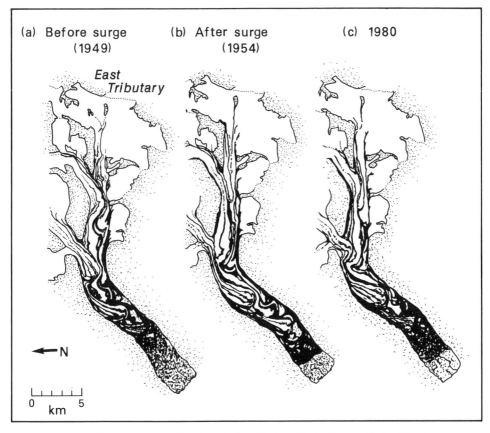

Figure 2.29   Evolution of moraines (black) on Susitna Glacier, Alaska Range, Alaska following a surge in 1951–2 (from T. S. Clarke 1991, with permission of the International Glaciological Society).

(e) large volumes of turbid meltwater are released;

(f) quiescent-phase ice structures, such as longitudinal foliation and medial moraines, are deformed into arcs and loops, resulting in the teardrop-shaped moraines that typify such glaciers (Fig. 2.27);

(g) widespread thrusting in the ice occurs at the surge front, a process associated with the passing through of a topographic bulge (Fig. 2.28).

In contrast, during quiescence, the crevasses ablate away and the ice becomes slow-moving or stagnant, although ice velocities gradually build up to a new surge in the reservoir area. In the ice-receiving area the ice ablates largely by down-wastage, and a drainage network with distinctive potholes commonly develops. A series of surge cycles may give rise to several sets of teardrop-shaped medial moraines, which give the glacier a complex and fascinating appearance that changes dramatically during each successive surge (Fig. 2.29).

2.9.2 Mechanisms of glacier surging

Although there is still much uncertainty as to the mechanisms of glacier surging, comprehensive investigations on a number of glaciers have led to the development of several hypotheses, and the dismissal of some of the more far-fetched ideas that were prevalent up to 15 years ago. The two hypotheses most favoured currently are dependent on a water-controlled triggering effect, whereby the subglacial drainage system is destroyed. However, one hypothesis, developed from studies on the temperate Variegated Glacier in Alaska, assumes a rigid substrate of bedrock; the other, from studies of the partially cold-based Trapridge Glacier in the Yukon, requires that the bed is deformable.

Rigid bed hypothesis
The rigid substrate model arises principally from the investigations made during the surge of Variegated Glacier in 1982–3, when measurements of borehole deformation revealed that *c.* 95% of ice motion was by basal sliding (Kamb et al. 1985). During quiescence, drainage at the base of the glacier, as in other non-surging temperate glaciers, takes place via a tunnel system, water emerging at the snout through a single portal. The prelude to the surge is the closure of the tunnel system, trapping meltwater and causing an increase in the basal water pressure. As a result, friction at the bed is lessened and the rate of basal sliding increases, allowing separation of the glacier from its bed and the formation of linked cavities controlled by bedrock irregularities. Surge velocities are attained once a linked cavity system has been established (Kamb 1987). When the linked cavity network develops into series of interconnected tunnels, water may be discharged more efficiently, often in the form of a flood, and the surge ceases. In reality, the surge of Variegated Glacier was irregular, with pauses and increases in velocity, the latter being referred to as mini-surges. The final cessation of the surge, in July 1983, was abrupt and was associated with a huge outburst flood of turbid water that drastically altered the morphology of the outwash plain (Kamb et al. 1985). By this time there had been wholesale transfer of ice towards the glacier snout, but the surge front ceased moving when within 1 km of the terminus, and the surge did not manifest itself in the form of a rapid advance of the glacier snout.

Deformable bed hypothesis
An alternative mode of surging has developed from the idea that ice may rest on a bed of soft sediment which is readily deformed (§2.6.3). Trapridge Glacier has been monitored continuously for over 20 years in the expectation that a surge is imminent, and it is thought that the glacier rests on a bed of deformable sediment which facilitates the initiation of a surge (Clarke et al. 1984). Unlike the Variegated Glacier, the Trapridge has a largely cold thermal regime, and in this respect is probably more typical of Arctic glaciers. The build up to a surge is marked by the development of a wave-like bulge in the lower reaches of the glacier. The bulge marks the boundary between warm-based ice up stream and cold-based ice down stream, and it has progressively moved down glacier during the period of observation at a rate of some 30 m yr^{-1}. As with the Variegated Glacier model, a surge in a glacier of the Trapridge type also begins with the destruction of the subglacial drainage system. However, in the latter case, the drain-

age system occurs within permeable sediments beneath the ice, in which, under certain circumstances, the effective permeability may be reduced. Progressive thickening of the ice leads to an increase in basal shear stress, which has the effect of reducing the permeability of till. Water pressure therefore builds up and the sediment is transformed into a slurry, the glacier then being able to flow at an enhanced velocity, which further weakens the sediment (Clarke et al. 1984). In this model, the surge terminates because the redistribution of ice leads to a lowering of the basal shear stresses to a level that allows the subglacial sediment to return to a permeable state. The Trapridge Glacier type model requires a large supply of readily erodible material to create a thick enough layer of subglacial sediment. Thus, soft bedrock and high rates of tectonic uplift may provide the necessary geological controls on the location of surge-type glaciers that behave in the manner envisaged for Trapridge Glacier.

Both the mechanisms discussed above seem plausible and they may not be mutually exclusive, even for the same glacier.

2.9.3 Do ice sheets surge?

From the standpoint of the stratigraphic record it is more important to know whether the Antarctic ice sheet is prone to surging, since it has been speculated that the Wisconsinan ice sheet may have behaved in this fashion (Clayton et al. 1985). Furthermore, it has been suggested that a major surge of the Antarctic ice sheet could so seriously disrupt ice/ocean interactions that Northern Hemisphere glaciation could once again be initiated (e.g. Hughes 1975). Theoretical evidence in favour of surging has been offered from the Lambert Glacier – Amery Ice Shelf system which drains about a fifth of the East Antarctic ice sheet (e.g. Budd & McInnes 1978, Allison 1979). However, this is countered by a variety of approaches involving the interpretation of the Landsat imagery of the glacier system (Robin 1979, McIntyre 1985, Hambrey & Dowdeswell 1994). More convincingly, some ice streams feeding into the Ross Ice Shelf from the West Antarctic ice sheet, have surge characteristics. Ice Stream B (underlain by a deformable bed) is currently flowing at more than $800\,myr^{-1}$ whereas its neighbour, Ice Stream C is practically stagnant with a velocity of only $5\,myr^{-1}$ (Whillans et al. 1987). However, buried surface crevasses in Ice Stream C, suggest a behaviour 250 years earlier that resembled that of B (Shabtaie & Bentley 1987). The contrasting behaviour thus inferred may be related to large-scale surging, but with a periodicity of hundreds of years. However, to date, there are no reports that any parts of the Antarctic ice sheet have, in fact, surged. The fact the cold ice caps with composite basins in Svalbard do surge (Solheim 1991), suggests that the phenomenon could occur on a larger scale in Antarctica.

2.10 Debris in transport

Glaciers act as conveyor belts for large volumes of eroded material. Unlike most other erosional depositional systems, the thickest sediments accumulate farthest from the source and, furthermore, even the most distal sediments contain coarse bouldery material. The manner in which sediment is transported out of the glacier system is largely dependent on the thermal regime. Cold, sliding glaciers carry a heavy basal debris load which is deposited as till. In contrast, highly dynamic temperate glaciers have relatively little basal debris but a high supraglacial load, most of which is modified by meltwater soon after deposition.

Following Boulton (1978), we can conveniently consider the movement of debris in relation to its transport path through the glacier. An understanding of these paths provides a better basis for interpreting the depositional processes and deriving the source areas of the sediment.

2.10.1 High-level transport

Debris carried at a high level in a glacier is derived from the following sources (partly after Drewry 1986):
- rockfall from adjacent mountain slopes;
- avalanche debris which includes snow, ice, soil and rock in various proportions from adjacent slopes;
- debris-flows from adjacent slopes;
- wind-blown dust (sand grade and finer);

Figure 2.30 In this photograph of Breithorn, Switzerland, steep tributary glaciers are bounded by rockfall-derived lateral moraines. As the glaciers enter the main trunk glacier, the Gornergletscher, flowing from left to right, the lateral moraines join to form medial moraines.

- volcanic ash;
- marine salts and microflora derived from sea spray;
- extraterrestrial, such as meteorites;
- pollutants from human sources;
- aerosols and gases.

Of these, rockfall is volumetrically by far the most important component on the surface of most valley glaciers. It tends to accumulate as **lateral moraines**, forming **medial moraines** where ice-flow units combine (Fig. 2.30). Commonly, these merge towards the snout, giving rise to a totally debris-covered surface. Rockfall debris is largely derived from frost-shattering processes and has a predominantly angular character (Fig. 2.31). Boulton (1978) has recorded the following characteristics, based on detailed examination of clast shape using Krumbein's (1941) sphericity/roundness chart, and grain-size distribution:

(a) clasts have low roundness (very angular and angular) and very variable sphericity (Fig. 1.4), and show little modification during transport;

(b) grain-size analyses indicate a predominance of clasts coarser than 1ϕ and a deficiency of fines, with little modification down-glacier.

Rock falling on to snow near the head of the glacier may be buried and may re-emerge below the equilibrium line. Rock falling on the surface near the equilibrium line will remain on the surface. In both cases the debris will be passively transported, unless it is able to reach a marginal or basal position by falling down crevasses or moulins.

Figure 2.31 Small medial moraine in the Glacier de Saleina, Switzerland, illustrating the characteristic nature of rockfall-derived debris. In the background is a serrated arête with a relatively low col formed as a result of breaching of the arête at the head of the glacier.

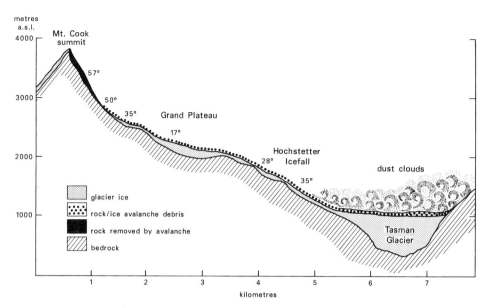

Figure 2.32 Rock and ice avalanche from Mount Cook, New Zealand that fell on to the Tasman Glacier, 14 December 1991 (after Chinn et al. 1992, with permission of the Institute of Geological and Nuclear Sciences, New Zealand).

Figure 2.33 Aerial view of the Mount Cook avalanche site (outlined) two months after the event that reduced the height of the mountain by 20 m. On the lower left is the Hochstetter Icefall already having absorbed much of the debris into crevasses. Below, the Tasman Glacier flows from left to lower right. The avalanche has obliterated the already debris-covered glacier surface and extended a considerable way up the lateral moraine on the far side.

Large-scale collapse of cliffs may turn free-falling rock into a fluid-like flow, especially if mixed with snow and ice, and the end product may be quite different from typical debris derived from rockfall. An impressive landslide took place in New Zealand in the early hours of 14 December 1991, when a buttress leading up to the summit of Mount Cook (3764 m) collapsed (Chinn et al. 1992; Figs 2.32 & 33). The rock and ice avalanche descended 2720 m, obliterating in its path the Hochstetter icefall and part of the lower Tasman Glacier. The lower edge of the avalanche rode up to a height of 70 m above the glacier surface, on the moraine wall that faces Mt Cook, 7.3 km from the source area. An estimated 14 million m^3 of rock and ice were involved, and being mixed with snow on the descent behaved as a fluid that may have reached a velocity of 600 km hr^{-1}. In the process, the height of Mount Cook, New Zealand's crowning peak, was reduced by 20 m. The resulting deposit was a pulverized mass of rock debris in which the originally angular shapes had been rounded, providing a marked contrast with material resulting only from freefall. Although many years may separate such events in alpine regions, individual glaciers may show signs of several major rockfalls, albeit on a somewhat smaller scale than that on Mount Cook.

Supraglacial streams may remobilize the rockfall material and sort it. Pockets of such debris accumulate in hollows and, when the channel is abandoned, the debris retards ablation, so leading to the development of dirt-cones.

2.10.2 Basal transport

Erosion at the bed of a glacier comprises various processes – crushing, fracturing and abrasion – which combine to produce a sediment very different from that carried on the surface. Debris eroded in this fashion is initially transported in a basal zone of traction, where particles frequently come into contact with the glacier bed and are retarded, so that large forces are imparted on both the particle and the bed.

Incorporation of debris into the basal ice is intimately associated with pressure melting and regelation (§2.7.2). Pressure melting on the upstream side of a bump is followed down stream by refreezing of the released water, thereby allowing a thin layer, a few millimetres to a few centimetres thick, to be added as regelation ice to the glacier sole. Any loose debris, ranging in size from clay to boulders, is incorporated into the regelation layer, and concentrations of debris may reach 50% or more by volume. Beneath a temperate glacier, regelation ice rarely builds up to more than a metre or so, as it soon melts again as the ice moves down-valley. In contrast, cold glaciers sliding on their beds build up dirty regelation layers (Fig. 2.34) as a succession that attains a thickness of several metres and persists down-valley. Regelation ice layers often appear foliated, like normal glacier ice, but derivation from water can often be determined using isotopic compositional contrasts (Souchez & Lorrain 1991).

A larger fragment (a cobble or boulder) may be incorporated into the glacier sole by ice deforming around it. It will be removed from its position, because shear within the basal ice results in rotation of the stone. Although a common process beneath sliding glaciers, it may also take place beneath cold-based ice to a limited extent.

Boulton (1978) demonstrated that clasts carried in basal ice (the zone of traction) have a

Figure 2.34 Debris-rich basal ice formed as a result of regelation; Taylor Glacier, Dry Valleys, Antarctica.

significantly higher roundness (predominantly subrounded and subangular) and slightly higher sphericity than those carried in high-level transport (Fig. 1.4). In addition, the boulders tend to have relatively smooth faceted surfaces, although sharp-edged fractures on otherwise smooth boulders are common. Striated abrasional facets are also common. Clasts that are flattish and do not roll readily attain typical flat-iron shapes with a strong preferred orientation of striae. Clasts that are more spherical and roll easily develop striations with many intersecting sets. Lithology has some bearing on whether a clast develops striae or facets.

Boulton (1978) also recorded that in basal debris the grain-size distribution is polymodal, but overall is depleted in the coarse fraction and enriched in the fine fraction in comparison with debris undergoing high-level transport. The high proportion of fines is the result of comminution of larger particles in the zone of traction, a process characteristic of a crushing mill, and one that produces the rock flour that gives glacial meltwater its milky appearance.

A further distinction can be made between the shape of stones carried in the basal zone of traction (and subsequently deposited as melt-out till) and that of boulders deeply embedded in lodgement till and subsequently modified by overriding ice. The latter often acquire a smooth striated, often bullet-nosed up-glacier termination and a sharply truncated down-glacier termination. The striae form a single, slightly diverging set. Overall, embedded clasts in lodgement till show an even greater degree of roundness than debris from the zone of traction (Fig. 1.4a,b).

2.10.3 Debris incorporated into an englacial position

A certain amount of debris reaches an englacial position from the bed of the glacier and may even emerge at the surface (Fig. 2.35), a process that operates in both warm and cold glaciers, but especially in the latter. Deformational processes in the ice responsible for this include thrusting and folding.

Figure 2.35 Debris-rich basal ice, thrust to the surface of the lower White Glacier, Axel Heiberg Island, Canadian Arctic. Note the predominantly subrounded and subangular character of the stones (cf. rockfall debris in Fig. 2.32).

Thrusting is a particularly common mechanism in cold glaciers, especially where a cold snout, frozen to the bed, is pushed from behind by sliding ice. Debris-rich basal ice may rise into the body of the glacier along a thrust. Many thrusts are related to transverse crevasse traces that have rotated into a orientation favourable for reactivation. Other planes of weakness may also promote thrust formation from the bed. Larger bodies of sediment may be incorporated into the ice where shear stresses within the glacier propagate, both into the underlying sediment and the sediment in front. The advancing glacier itself shows prominent thrusts and shear zones and the author suggests these may extend downwards on to the frozen sediment (Fig. 2.36). A **sole thrust** (in the manner of deformation in rocks in mountain belts) may develop in the sediment beneath, and in front of, the ice. As the ice advances, new thrusts develop progressively more forwards than the previous ones. Structural geological concepts have been applied to thrusting in push moraines by Croot (1987), but they apply equally well to the processes going on in the ice itself.

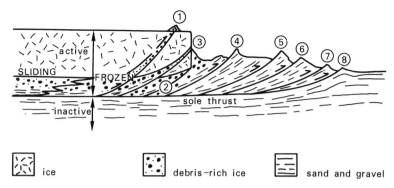

ice   debris–rich ice   sand and gravel

Figure 2.36 Sketch to show how large masses of frozen subglacial sediment may be incorporated into an englacial position by thrusting. The thrusts extend down through the ice into the frozen sediment, there being no sliding at the ice/sediment interface. Forward movement of the ice leads to the development of a "sole thrust" from which upward-curving subsidiary thrusts develop progressively in front of each other (numbered sequentially, 1–8).

A second tectonic mechanism of debris incorporation involves folding of dirty basal ice. The folds tend to have limbs that are subparallel to the bed, and the hinges are perpendicular to the direction of ice-flow. Hudleston (1976) has investigated such folds in the Barnes Ice Cap and has demonstrated that they can form only when perturbations in flow occur. Repeated folding may cause the regelation layer to rise to a high level within the glacier.

The transfer of debris from the surface to an englacial position is associated with the opening of crevasses, a process which may be very effective in the short term (several years). With reference to the Mount Cook landslide referred to above, the crevasses in Hochstetter icefall had, within just a few weeks, swallowed almost all the surface debris that had initially obliterated the glacier surface (Fig. 2.33). Supraglacial streams may also be effective in transporting debris into the body of the ice, especially when they enter a moulin.

3 Glacial erosional processes and landforms

3.1 Introduction

Glacial erosion has its most profound effect scenically in mountain environments (Fig. 3.1), where material stripped off is rapidly transported to lower lying regions. However, glacially eroded lowland areas are even more extensive. Large areas of the Laurentian Shield of North America, and those parts of the Sahara influenced by Ordovician glaciation, are examples of lowland areas exhibiting significant glacial erosion. Estimates of the amounts of erosion vary by orders of magnitude but, judging by the column of sediment transported by glaciers offshore, must reach several tens or hundreds of metres.

3.2 Processes of glacial erosion

Glacial erosion comprises several processes: abrasion, fracture of structurally homogeneous rock, and fracture of jointed rock, each followed by entrainment of debris into the basal zone of the glacier, and meltwater erosion. Frost shattering of valley-side cliffs, steepened by glacial erosion and the removal of debris supraglacially, must also be taken into account.

Abrasion involves ice containing debris overriding bedrock. The debris scores the bedrock to produce fine-grained material (grain size generally < 100 µm), as well as generating striations and other features discussed below (§3.3.2). On the basis of laboratory work and direct observations beneath glaciers, the principal factors affecting glacial abrasion appear to be the presence and concentration of basal debris, the sliding velocity of the glacier on its bed, and the transport of debris towards bedrock, so continually renewing the abrasive surface. Other factors affecting the rate and type of abrasion include:

- ice thickness (generally the thicker the glacier the greater the pressure exerted on the bed);
- the presence of water at the glacier base (high water pressure has the effect of buoying up the glacier, so reducing the effective pressure, although this is counteracted by increased basal sliding);
- the relative hardness of basal debris in transport compared with bedrock;

81

Figure 3.1 The Matterhorn, viewed from Allalinhorn, Switzerland. This classic horn has been eroded by glaciers on three sides. Two of these glaciers are visible here, divided by the steep Hörnli Grat, the arête by which the first ascent of the mountain was made (photograph courtesy of Jürg Alean).

- the size and shape of particles embedded in the ice;
- the effectiveness of removal of the eroded debris, especially by meltwater.

The relative importance of these factors varies according to the thermal regime of the glaciers. Cold glaciers, especially those frozen to the bed, are less effective agents of erosion than those that are sliding. Thus, the least active glaciers tend to occur in polar regions.

Under certain circumstances, ice (or debris in the ice) exerts sufficient force on homogeneous bedrock to cause it to fracture. Crescentic fractures are one indication of this, apparently sheared boulders are another. Fracture can also occur by **pressure release** after the glacier has disappeared (Lewis 1954). In effect, after rapid erosion, newly exposed rock is in a stressed condition, and joints may develop, commonly resulting in **exfoliation** of large sheets of rock on steep valley sides. Additionally, subaerial freeze/thaw in the presence of meltwater promotes fracturing, after the development of jointing in response to **dilation** (Harland 1957).

Rocks that are already jointed before the arrival of a glacier are particularly susceptible to erosion. Stratified, foliated and faulted rocks are also likely to be exploited more than structureless rock. The joints may or may not have been exploited by weathering before the onset of glaciation; if not, freeze/thaw beneath a glacier, the base of which is close to the pressure melting point, is a potential mechanism for loosening the blocks. The down-glacier sides of **roches moutonnées** are a favourable locality for block removal (Carol 1947). Once loosened, large blocks (**rafts**) may be detached from the bed.

Once material has been detached from the bed, it may be incorporated into the basal zone of the glacier by one of two processes. Small particles are incorporated by regelation, especially on the downstream side of bedrock obstacles, while larger blocks are picked up by the ice deforming around and surrounding them. For a glacier of simple shape, the efficacy of glacial erosion and transport reaches a peak near the equilibrium line.

3.3 Landforms resulting from erosion by glaciers

3.3.1 Modes of erosion

The most dramatic effects of direct glacial activity are erosional, and are represented by such major landforms as glacial troughs, cirques and alpine forms (horns and arêtes), medium-scale features, such as roches moutonnées, and small-scale features, such as polished and striated pavements, with their associated markings. Excellent detailed descriptions of these features, with many examples, have been given by Embleton & King (1975), Sugden & John (1976) and Prest (1983).

As ice flows over an obstacle it tends to abrade an up-glacier-facing surface, thereby smoothing it. In contrast, surfaces that face down-glacier are subject to several mechanisms, including initial bedrock fracturing, loosening and displacement of rock fragments, and incorporation of those fragments into the base of the sliding glacier; such surfaces are rough, and are commonly referred to as plucked or quarried. Drewry (1986) has provided an extended quantitative treatment of these processes.

Features that are solely the result of abrasion are referred to as streamlined, and they include lowland bedrock areas that have been scoured, glacial troughs, watershed breaches, domes, whaleback forms, grooves, p-forms, striations and polished surfaces. Other forms, which are the result of both abrasion and rock fracture, include trough-heads, cirques, cols, roches moutonnées and riegels; these are only partly streamlined. Another group of small-scale features are the result of rock crushing, whereby stones embedded in the base of moving ice subject the bed to sporadic impact; these are referred to as friction cracks and are non-streamlined forms. A final group of landforms are in effect residual, reflecting what is left after a combination of abrasion and fracturing by ice, and frost-shattering and hillslope movement, has attacked an elevated mass of rock. Such landforms are characterized by horns, arêtes and nunataks.

A landform classification based on that provided by Sugden & John (1976: 169), but modified according to the processes discussed by Drewry (1986), is given in Table 3.1. The following discussion treats each group in turn, more or less according to scale, with the largest first.

Table 3.1 Classification of glacial erosional landforms according to process of formation, relief form and scale.

Process	Relief type	Landform	Scale (approx. range)
Abrasion by glacier ice dominant	Streamlined	Areal scouring	~10 km – 1000 km
		Glaciated valley	~1 km – 100 km
		Fjord	~1 km – 100 km
		Hanging valley	~1 km – 10 km
		Watershed breach	~1 km – 10 km
		Whaleback	~100 m – 1 km
		Groove	~1 cm – 1 km
		P-forms	~1 m – 100 m
		Striations	~0.1 mm – 1 cm
		Polished surfaces	~0.1 mm – 1 km
Combination of abrasion and rock fracturing by glacier ice	Part streamlined	Trough head	~1 km – 10 km
		Rock step	~1 km – 10 km
		Cirque	~1 km – 100 km
		Col	~1 km – 10 km
		Roche moutonnée	~100 m – 1 km
		Riegel	~1 km – 10 km
Rock crushing	Non-streamlined	Lunate fracture	~1 cm – 10 cm
		Crescentic gouge	~1 cm – 10 cm
		Crescentic fracture	~1 cm – 10 cm
		Chattermark	~1 cm – 10 cm
Erosion by glacier ice and frost shattering	Residual	Arête	~10 km – 100 km
		Horn	~10 km – 100 km
		Nunatak	~1 km – 1000 km

Scale axis: 0.1 mm 1 mm 1 cm 10 cm 1 m 10 m 100 m 1 km 10 km 100 km 1000 km

3.3.2 Glacial abrasion

Areal scouring

Some lowland areas bear abundant signs of glacial erosion on a regional scale, such as large parts of the Laurentian and Baltic shields. This is referred to as **areal scouring**. A low amplitude but irregular relief, sometimes referred to as "knock and lochan" topography after the heavily scoured landscapes of the North-West Highlands of Scotland, characterizes such terrain. In this area, ice masses flowed over, and scoured, Lewisian basement gneisses, occasionally leaving the summits of the younger Precambrian Torridonian Sandstone peaks as isolated nunataks. Bedrock is exposed nearly everywhere, with roches moutonnées (see below) or rocky knolls (knocks) and lake (lochan)-filled depressions. Structural features, such as joints, faults, dykes and steeply inclined strata, were selectively exploited by the ice. This type of landscape, which has a relief of about 100 m, must have been formed beneath a sliding ice sheet. Such conditions probably exist beneath large parts of the Antarctic and Greenland ice sheets today.

Sometimes associated with areally scoured landscapes are sets of parallel grooves and **bedrock flutes**, which in some respects resemble fluted moraine in predominantly depositional environments (Ch. 4). Genetically they may be linked, in the sense that obstacles to flow may initiate a flute. Good examples are to be found eroded in the Precambrian basement gneisses

Figure 3.2 A sandstone tor of periglacial origin in the Kennar Valley area, Victoria Land, Antarctica. This area was once covered by ice as indicated by scattered erratics, but there was little erosion, probably because the ice was frozen to the bed. In contrast, deep erosion took place in the background where an ice stream was sliding on its bed.

Figure 3.3 Contrasting forms of glaciated valleys. (a) A true U-shaped profile, the Yosemite Valley in California. The near-vertical crag on the left is El Capitan, 1500 m high. (b) A more typical glaciated valley, Glen Rosa, Isle of Arran, Scotland, with a more open (parabolic) cross section. Note the small breached watershed, cirque and horn named Cir Mhòr at the head of the valley.

of the Laurentian (Canadian) Shield, notably in northeastern Alberta (Embleton & King 1975).

In contrast to these areally scoured regions, some areas of low or moderate elevation may have experienced little glacial erosion, even though inundated by an ice sheet. In such cases only isolated erratics on a deeply weathered surface with fluvial or periglacial forms will indicate the former presence of ice. Ice-free areas in Antarctica illustrate well the effectiveness of glacial erosion where ice is channelled into valleys, but also its rather protective nature on the intervening plateau remnants (Fig. 3.2), even though erratics indicate that ice covered the entire region. It is likely that the thinner ice over the plateau was frozen to the bed and slow-moving, protecting it from erosion. Much of the present-day ice cover, too, is frozen to the bed, and erosion is minimal.

Glaciated valleys

Glacial troughs, a term which embraces both valleys and fjords, are probably the most spectacular manifestations of glacial erosion. They are predominantly the product of abrasion as a result of ice being strongly channelized, but rock fracturing and plucking down stream of smoothed faces is common until irregularities have been removed. It is widely recognized (Flint 1971), however, that most troughs follow the line of existing river valleys, but the glaciers act as over-deepening, widening and straightening agents.

Many glacial troughs are eroded by valley glaciers descending from high mountains, such as where cirque glaciers extend towards lower ground, as in the European and New Zealand Alps, the Western Cordillera of North America, the Himalaya and other temperate areas today. Other glacial troughs develop beneath, or extend from, ice sheets or ice caps, where ice-streaming is prominent, as beneath the Greenland and Antarctic ice sheets, and other Arctic or sub-Arctic ice caps. In the case of East Greenland, the landscape is characterized by deep incisions into plateau-like terrain with striations to the top of the cliffs, while the plateaux themselves bear little evidence of erosion. The incisions form a complex dendritic pattern, depending on the alignment of preglacial river valleys, or a "chocolate tablet-like" arrangement, if controlled by intersecting bedrock faults. Modification involves over-deepening, the development of trough-heads, steps, riegels and basins, the steepening of the valley sides and truncated spurs.

In cross section, glaciated valleys are often described as U-shaped. Well known valleys which approach a strictly U-shaped profile include Lauterbrunnental in Switzerland and the Yosemite Valley in California (Fig. 3.3a). These valleys feature flat bottoms, reflecting infilling by hundreds of metres of glaciofluvial gravels or lake sediments. The lower rock and scree slopes end abruptly against this flat floor. However, truly U-shaped profiles are exceptional, and most glaciated valleys have much less steep sides (Fig. 3.3b). An "ideal" glaciated valley, in mathematical terms, has a parabolic profile, according to the formula:

$$y = ax^b$$

where the x and y axes correspond, respectively, to the depth of the valley and half the width. Most ideal troughs have an exponent (b) of about 2. On the other hand, many troughs have asymmetrical profiles commonly related to differences in bedrock hardness. In the English Lake District, the Buttermere Valley has a steep southern and southwestern flank comprising rocks of the hard Borrowdale Volcanic Group and intrusive syenite, whereas the opposite flank, with its gentler, less craggy aspect, is made of softer mudstones of the Skiddaw Group. The steeper reaches of many alpine-type glaciated valleys are characterized by a V-shaped profile. This may result more from vigorous erosion by subglacial meltwater under high pressure than from abrasion by ice. Even many parabolic profiles are interrupted by V-shaped notches in the valley floor.

In contrast to the normally smooth cross sections, the long profiles of glaciated valleys are extremely irregular, as a result of uneven over-deepening by the ice. Over-deepening is controlled by many factors, the most important of which is the presence of constrictions in the valley or the entering of a tributary glacier into the main stream, both of which result in accelerated ice-flow and enhanced erosion; also important are changes in bedrock geology and structure. As a result, many glaciated valleys have multiple rock basins filled by lakes. Often these valleys (and fjords) are over-deepened below the general level of the continental shelves offshore. Many examples exist in western Scotland; Britain's deepest lake, Loch Morar, is 315 m deep but its surface is only 15 m above sea level. Some of the largest ice-scoured hollows, now filled by lakes, lie near the perimeter of the Alps, the deepest being Lago di Como (410 m) in northern Italy. Many of these lakes are deepened further by being dammed by material scooped out of the hollows and dumped as end moraines. Most glaciated valley lakes begin to fill rapidly with glaciolacustrine and glaciofluvial material, but they become much more stable when this source is cut off. Gravel fans from a side valley may grow out into a lake, while suspended sediments will also gradually fill the lake (Ch. 6).

In the geological record, reports of glaciated valleys are rare, but a few examples have been described from the Late Palaeozoic sequences in Gondwanaland (South Africa and Antarctica).

Fjords

The term **fjord** is of Norwegian origin (spelt **fiord** in North America and New Zealand) and, in a morphological sense, refers to a glacial trough, the floor of which is below sea level. Fjords still occupied today by glaciers are found in Antarctica, Greenland, the Canadian Arctic Islands, Svalbard, Novaya Zemlya, southern Alaska, Chile and South Georgia. Fjord coastlines from earlier glacial phases are well developed also in Norway, Scotland, Iceland, British Columbia and New Zealand.

Genetically, fjords and glaciated valleys are similar, and frequently pass into one another. They possess many common characteristics, of which the most striking is the depth to which they have been over-deepened, illustrating the effectiveness with which glaciers can erode, irrespective of the level of the sea.

Like glaciated valleys, fjords have a parabolic profile, although many more of them approach a true U-shape with vertical sides, particularly where they are cut in hard crystalline rocks beneath former ice sheets, such as in Greenland, southwestern Norway, Baffin Island, South Island of New Zealand (Fig. 3.4). Greenland and Norway have particularly long fjords, many over 100 km long, and the depth of some, from crest to sea bottom, attains 3000 m. Probably the world's biggest fjord is the Lambert Graben in East Antarctica, but it is filled by an ice stream that drains a large part of the East Antarctic ice sheet. If this ice were to disappear a

(a)

(b)

Figure 3.4 Fjords in Greenland and New Zealand. (a) The world's longest open fjord, Nordvestfjord, which, with Scoresby Sund, stretches inland for 300 km from the open sea. An active tidewater glacier at the head (cf. Fig. 2.8) discharges tabular icebergs which can be seen heading towards the open sea. Rotten (greyish) sea ice floats on the fjord surface in the foreground. (b) The head of New Zealand's best known fjord, Milford Sound, with the horn, Mitre Peak (1692 m) on the left. Glaciers last occupied this fjord in the late Pleistocene epoch, although mountain glaciers (right) still survive. Successive phases of down-cutting are evident from the change in slope of the fjord walls.

fjord *c.* 800 km long, 50 km wide and 3 km from crest to floor would become exposed. These exceptional dimensions reflect the far longer period of glacierization in East Antarctica (at least 40 Ma) than in other parts of the world, as well as the softer bedrock that occurs on the floor of the Lambert Graben compared with the adjacent land.

Many fjords, especially those in Norway, have a characteristic long profile which reflects a greater degree of glacial erosion at their inner ends, as it is there that the ice in contact with the bed is at its most dynamic. Thus, fjords deepen quickly from their headward ends and then progressively shallow seawards, although this general trend is often interrupted by the presence of distinct basins. The mouth of a fjord is often marked by a **sill** with shallows, rocky reefs or even islands. The best known example is Norway's Sognefjord, which has a maximum depth of 1308 m, but at the entrance the fjord is only 3 km wide and 200 m deep. The sills may be capped by morainic debris, although the depth of this material in relation to the eroded basins is small.

Recorded fjord depths may only partly reflect the depth to which glacial erosion has occurred, since large volumes of sediment often fill the over-deepened basins, sometimes to depths of several hundred metres, as in Glacier Bay, southern Alaska today (Ch. 7).

Less spectacular, but none the less renowned for their scenic beauty, are the sea lochs of western Scotland. Although with gentler slopes and a relief rarely exceeding 1200 m, these fjords show many of the same characteristics of the larger fjords, namely multiple basins, deeper inner parts and rocky sills. In many cases, such as Loch Morar and Loch Maree, the sills have been raised above sea level since deglaciation as a result of isostatic rebound, creating freshwater lochs separated from the sea by short rapids.

Some fjords show progressive stages of down-cutting, and even the upper part of the former river valley profile may be preserved. A "U" within a "U" may be also apparent (Fig. 3.4b). Major landforms associated with fjords include trough-heads, hanging valleys and breached watersheds (see below).

Fjord systems are as varied in terms of their spatial arrangement as glaciated valley systems. Some have linear trends related to faults, as in East Greenland and Scotland. Other fjord systems that were fed by several equally important glaciers, such as Glacier Bay in Alaska, have a dendritic arrangement. Fjord systems with sinuous branches are common in south-western Norway, although in detail the fjord walls are smooth, and only the spurs truncated.

Breached watersheds

The action of a valley glacier, when it grows sufficiently to spill out of its constraining trough as a diffluent ice stream, leads to downward abrasion of the lower cols. These cols may themselves attain a parabolic form in cross section and eventually be so deeply eroded that the main trunk glacier is diverted. It may be, however, that the effect of **diffluence** of ice over a col is simply the transfer of some ice from one valley to an adjacent one.

Most glaciated mountain regions demonstrate glacial diffluence. The British Isles have many examples, some of which are well described by Embleton & King (1975). A good example connects Glen Sannox and Glen Rosa on the Isle of Arran (Fig. 3.3b). The former, with its own valley glacier, was partially blocked at its mouth by a major glacier from the Grampian Highlands flowing south down the Firth of Clyde. As a result some of the ice accumulating at the head of Glen Rosa spilled south over a col into Glen Rosa, as indicated by abraded rocks on the col and its over-deepened character. Glacial diffluence on a widespread scale is evident in the Grampian Highlands. A well known example is the Lairigh Ghru, a high pass that divides the 1300–1400 m high Cairngorm Plateau into two distinct parts. Here ice in the upper Dee Valley was restricted by ice to the south, and thus overspilled the col to the north, lowering the col by an estimated 230–840 m. Many other examples in Scotland, Wales and the Lake District have been reported.

Watershed breaching is also a feature of fjord landscapes, and notable examples occur in southwestern Norway and East Greenland (Fig. 3.5). The latter area provides several excellent examples of watersheds having been cut down to near the level of the base of the main trunk glacier.

When diffluence attains a level at which all cols are being used to discharge ice the term glacial **transfluence** is used. The process is well illustrated by past and present ice sheets, and frequently represents a shift in the ice divide inland of the original mountain divide. This happened in Scandinavia and Scotland, as revealed by the transport of erratic boulders eastwards up and over the respective mountain divides. The Antarctic demonstrates transfluence on a vast scale. Here, the East Antarctic ice sheet impinges against the Transantarctic Mountains at a level of around 4000 m and discharges ice towards the Ross Ice Shelf over buried cols. Some of the biggest and fastest-flowing ice streams in Antarctica are cutting through

Figure 3.5 A near U-shaped breached watershed in East Greenland, perched high on a precipitous rocky ridge. The main glacier flowed down Nordvestfjord, behind the ridge from left to right, and some ice spilled over this col joining another ice stream that had overridden a low-level col to the left.

(a)

(b)

Figure 3.6 Hanging valleys. (a) The Steall Waterfall forms a graceful cascade as it drops into Upper Glen Nevis, Scotland. (b) This hanging valley terminates at the vertical northern flank of Milford Sound, New Zealand, creating a fine waterfall.

the Transantarctic Mountains, including the Beardmore and Axel Heiberg glaciers which, respectively, provided routes to the South Pole for Scott and Amundsen in 1911–12. Many parts of the Greenland ice sheet are behaving in a similar fashion. Greenland is more-or-less fringed by mountains which, even though over 3000 m high in places, are only partly able to constrain the ice sheet, and many glaciers spill out seawards through breaches in the coastal mountains. Another area of extensive ice cover, the Icefield Ranges of Alaska and the Yukon, demonstrates the same process of watershed breaching, as ice escapes seawards to the south, through ranges over 4000 m high.

Hanging valleys

Hanging valleys are a characteristic feature of glaciated mountain landscapes, although not exclusively confined to them. The hanging valleys were occupied by tributary glaciers that were not as effective at eroding downwards as the main trunk glaciers. Thus, one finds glacially shaped side valleys ending abruptly against the steep wall of the valley (Fig. 3.6) or fjord (Fig. 3.7), and this provides the site for waterfalls to form. Cirques are a type of hanging valley, but normally the term is restricted to side valleys that had valley glaciers in their own right. One of the most famous examples of a hanging valley landscape is the Yosemite of California, where the Ribbon and Bridalveil falls both drop several hundred metres over vertical cliffs into the main valley (Fig. 3.3a). Most fjord regions have fine hanging valleys, as do high-alpine regions. Even relatively low highland areas such as those of Britain have well developed hanging valleys, although seldom do they have sheer rock faces below.

Domes and whaleback forms

If a glacier meets an obstruction, it may not be able to abrade it totally, but may leave it as an upstanding, smoothed rock hillock. There are two types, **domes** and **whaleback forms** which are totally the product of abrasion. On a similar scale, roches moutonnées (§3.3.3) represent a combination of abrasion and plucking.

Domes are relatively unusual features and are best developed in areas of homogeneous bedrock which, after being eroded by the ice, is subject to **exfoliation**. This process involves peeling off curved slabs of rock like the skin of an orange, so presenting a smooth, streamlined form to the next ice-re-advance. Among the finest examples of glacially eroded domes are those cut into granite in the Yosemite National Park. Some of them are nearly symmetrical and reach heights of several hundred metres (Fig. 3.7).

Whaleback forms, which have also been referred to as rock drumlins, are a good example of a streamlined landform, and are typically tens to a few hundred metres long, and from less than one to tens of metres high. Length:width ratios are commonly 1:2 to 1:4, and they are orientated parallel to flow. In this respect they are similar in shape and orientation to drumlins made of drift and, indeed, they may occur as peripheral members of drumlin fields. Clear examples of streamlined whaleback forms are rare, but Embleton & King (1975) and Sugden & John (1976) have provided examples from Scotland, Iceland, and the Baltic and Canadian shields.

On a larger scale, spurs and interfluves may be tapered and smoothed. The Finger Lakes of New York State are an example. All these forms carry the small-scale markings described below.

Figure 3.7 A series of ice-abraded domes of granite rising above the glacial Tenaya Lake basin, Yosemite National Park, California. Ice-transported granite boulders litter the shore and shallows of the lake.

Striated, polished and grooved surfaces

Striations are among the most common features of glacial erosion. They are finely cut, U-shaped grooves on the surface of bedrock that has been scored by stones in the base of a sliding glacier (Fig. 3.8a). Individual lines are sometimes a metre or more long, occurring in parallel. Glacial striations are quite varied in form. If the cutting tool is rotated, so presenting a new cutting edge, the resulting striations appear to step sideways or lie *en echelon*. Many striations are asymmetrical, blunt and deep at one end, tapering at the other end (Fig. 3.8b). These are sometimes known as **nail-head** striations and they usually indicate ice-flow away from the tapering end, but, on surfaces sloping up-glacier, gouging may give rise to the opposite pattern. Some striated surfaces have **rat-tails** (Fig. 3.8c), which are minor ridges extending down stream from knobs of more resistant rock that have protected the more easily striated material on either side. Although uncommon, rat-tails are good indicators of the sense of direction of ice-flow as opposed simply to its orientation.

Striations commonly show a wide range of orientations in a small area. On uneven bedrock they reflect all the irregularities of the basal flow of the glacier. Even on flat surfaces, rotation of the gouging stones results in marked deviations from the mean flow direction, while different recognizable sets may reflect longer term changes in the directions of flow. There-

94

Figure 3.8 Small-scale features of glacial abrasion. (a) Well preserved striations made by Late Palaeozoic glaciers on Early Proterozoic dolomite, near Douglas, Karoo, South Africa. Although these striations display strong parallelism, a curving set of "tram-lines" indicates occasional variations in ice-flow direction. Note the diamictite of the Dwyka Formation at top right. (b) Short nail-head striations on a steep rock face that flanked the McBride Glacier, Glacier Bay, Alaska in the 1970s. Ice-flow was towards the right; note how the striations score progressively deeper into the bedrock until the abrading tool flips out. These nail-heads formed at a later stage of erosion than the dominant gently inclined set of regular striae. (c) Rat-tails in dolomite, developed down-glacier of resistant chert concretions which protect the bedrock from erosion (Wordie Gletscher, East Greenland). Ice-flow was from bottom to top of the picture. (d) Ice-abraded boulder pavement at the base of the Late Palaeozoic Dwyka Tillite, Elandsvei, near Tweifontein, Karoo Basin, South Africa. Note the regular nature of the striations, formed as the ice flowed from right to left.

fore, it might be thought that striations are of little use in reconstructing the mean directions of movement of former glaciers and ice sheets. Nevertheless, such directions have been obtained successfully in many cases. The measurement of a large number of striations in a small area facilitates statistical treatment and a graphical indication of mean ice-flow (especially if supported by other evidence), even if individual striations are orientated up to 90° from the mean.

Minute scratches in large quantities give rise to a polished appearance of the rock, although they may be visible only under a magnifying glass. The degree of polishing depends on the fineness of the abrading material. Fine-grained rocks with little bedding, foliation or jointing, are the most suitable for the generation and preservation of polished surfaces. Coarsely crystalline and easily weathered rocks are the least likely to bear well preserved striations and polished surfaces. Exceptionally hard rocks, such as quartzites, are also unlikely to acquire many striations. Sometimes, till itself may be subjected to glacial erosion, resulting in bevelling of the stones embedded in the till and the development of a striated boulder pavement (Fig. 3.8d). Even boulders in glaciomarine sediment may become striated if they are overrun by grounded ice (§8.5.6). On exposure to air, striations and polish are often soon lost through weathering. They tend to survive best if subsequently buried rapidly by till. In this way, striations dating from 10000–20000 years ago may be well preserved, as in the highland areas of Britain. In the older geological record, striated surfaces are often exposed beneath tillite horizons and provide convincing evidence of former glaciation. Impressive exhumed striated pavements of Late Palaeozoic age are widespread in southern Africa (Fig. 3.8a) and other Gondwana continents. Ordovician and Neoproterozoic pavements are found in the Sahara; others of Neoproterozoic age are known from places as far apart as China and Norway.

Another type of striation arises when ice flows over bedrock that has a thin film of mud or sand. Such striations are recognizable by **plough-marks** associated with stones pushing the soft sediment (Fig. 3.9a), and by slumping of the sediment into the groove (Fig. 3.9b).

As indicators of glaciation, striations have certain limitations. They are insufficient in themselves since other non-glacial agencies can give rise to striations. For example, floating lake, river or sea ice with embedded stones is capable of scratching a rock, although the resulting striations tend to be shorter, more irregular in orientation, and are found only in limited areas.

Debris-flows, both subaerial and submarine, can striate the surfaces over which they are flowing; volcanic debris-flows are particularly capable of this. Avalanches can also scratch a surface. Tectonic lineations sometimes closely resemble glacial striations. All these types of striations, however, tend to show a greater tendency to parallelism than do those formed by glaciers.

(a) (b)

Figure 3.9 Late Palaeozoic striations and related features formed in a thin veneer of soft sediment that lay on quartzite bedrock, Elandsvei, South Africa. (a) Ploughed ridge in front of pebble (above coin) with groove behind; flow was from right to left. (b) Flowage of soft sediment from top to bottom of picture has created small lobate features that partially obliterate the groove extending to the right of the head of the hammer.

Striations are gradational with large furrows or **grooves** carved not only out of soft sediment-ary rocks but also out of granites and gneisses. Grooves themselves may even be regarded as gradational into glacial valleys. Many grooves appear to be simply enlargements of single striations. Ice may occupy such grooves and further result in localized abrasion. Grooves attain depths and widths of a few metres, and lengths of several hundred metres (Fig. 3.10a,b). Many have overhanging walls, which themselves are striated. Spectacular examples on islands

(a)

(b)

Figure 3.10 Grooves resulting from prolonged glacial abrasion. (a) Steep-sided groove, measuring about 3 m deep and 8 m across formed under the Quaternary Laurentide ice sheet, Whitefish Falls, near Sudbury, Ontario. (b) Open, wave-like grooves formed under the Late Palaeozoic ice sheet that covered Gondwanaland, near Douglas, Karoo, South Africa. Note that in both cases the grooves themselves are striated.

in western Lake Erie have been described by Goldthwait (1979) and grooves 30 m deep and 1.5 km long have been reported in the Mackenzie River Valley in Canada (Flint 1971: 89).

Plastically moulded (or p-) forms

Some glaciated rock surfaces exhibit complex, smooth forms **p-forms**, features which have been the subject of considerable argument. Various origins have been proposed for these enigmatic features: (a) the normal processes of glacial abrasion, (b) the movement of saturated till at the base or sides of a glacier, and (c) the action of meltwater, especially under the high pressures that may exist in places beneath a glacier. All three processes may in fact be responsible for the different features within this group. The best preserved p-forms have been described from areas of resistant igneous and high-grade metamorphic terrains. Norway is particularly well endowed with p-forms, where they occur along deeply incised glacial valleys and fjords. p-forms are also to be found on some of the western isles of Scotland, e.g. Mull and the Garvellachs.

On the basis of work in northern Norway, Dahl (1965) classified p-forms according to their geometry. Those that are likely to be more the product of glacial abrasion than any other process are **Cavetto forms** and **grooves**.

– **Cavetto forms** are channels cut on steep rock faces and orientated parallel to the valley sides. They may be up to 0.5 m deep, have sharp edges, and the upper part may overhang. Striations and crescentic gouges occur within them. Such linear features may well be the products of abrasion.
– **Grooves** occur on flat open surfaces (as mentioned above), but have rounded edges. Often cut in extremely hard rocks, they are also likely to be the result of glacial abrasion.

Other p-forms include **Sichelwannen**, **curved** and **winding channels** (alternatively known as **Nye channels**. These are described in §5.3.1 as the predominant process is likely to be meltwater erosion.

3.3.3 Abrasion and rock fracture combined

Trough-heads and valley-steps

Many glaciated valleys and fjords terminate abruptly inland at steep, rocky faces called **trough-heads** or **trough-ends**. They mark the position where over-deepening in the longitudinal profile has occurred, but their origin is obscure. It has been suggested (Sugden & John 1976: 184) that a trough-head represents a switch from sheet-flow of ice (as at the edge of an ice cap) to channelized flow, combined with a change in the basal ice condition to one that promotes slip. However, trough-heads are morphologically similar to **valley-steps** and the latter occur within alpine valleys where ice-flow is channelized throughout.

The trough-head or valley-step has a heavily "plucked" appearance, but ice-abrasion marks may be found on the less steep, down-valley-sloping surfaces on the craggy face itself, as well as on the crest of the trough-head. It is probable that some trough-heads and valley-steps are related to original breaks in slope caused by the cropping out of harder rock. As the ice flows over a vertical face it loses contact with the bed and creates a cavity. Here, freezing and thawing processes may assist in the loosening of blocks. The ice regains contact with the bed lower down and abrades it, and the process is repeated down the cliff. The combination of these processes may help to perpetuate these features, but insufficient work has been done on their morphology to characterize adequately the processes responsible for their formation.

Some of the most spectacular trough-heads occur at the heads of fjords in southwestern Norway. Several branches of Sognefjord have good examples, the best known being at the head of Aurlandsfjord where a branch of the Oslo–Bergen railway spirals down about 700 m altitude from Myrdal to the village of Flåm on a short sediment fan close to fjord level. Another impressive example lies at the head of Gasterntal in the Berner Oberland of Switzerland (Fig. 3.11). In Britain, fine trough-heads, but on a smaller scale, occur at the head of the Buttermere Valley in the Lake District and at the head of Loch Avon in the Cairngorms. In the USA, the Yosemite Valley terminates in a trough-head.

Valley-steps are more common than trough-heads. Many valleys in the Alps have several rock-steps, each of which occurs where the valley narrows. In Wales, a series of valley-steps originating in a cirque occur on Snowdon, and there is another fine example in the Cuillins of Skye in Scotland.

Both trough-heads and valley-steps have been likened to large-scale roches moutonnées (below) and it is probable that all these forms are the result of common processes.

Figure 3.11 Trough-head in Gasterntal, Berner Oberland, Switzerland. The snout of the glacier Kanderfirn rests at the top of the heavily plucked rock face and the Little Ice Age lateral moraine is visible on the right.

Figure 3.12   Riegel or transverse rock barrier breached by the river at the right. The ice in the foreground is Franz Josef Glacier, Southern Alps, New Zealand.

Riegels

Riegel is a German term given to a rock barrier that extends right across a valley, either holding back a lake or, when breached, as an upstanding transverse ridge (Fig. 3.12). The barrier may show limited signs of abrasion on the up-glacier and much fracturing on the down-glacier side. Riegels usually form where a band of resistant rock crosses a valley.

Cirques

Of all the landforms of glacial erosion, **cirques** are among the most fascinating, and they have long been regarded as one of the surest indicators of past glacial activity. Although the term cirque embraces a broad family of landforms, in its most characteristic form it resembles an armchair-shaped hollow high up on the mountainside (Fig. 3.13). The widespread occurrence of cirques (the term is French, but used internationally), is reflected in a wide variety of local names. In Britain the term **corrie** (from the Gaelic, *coire*) is normally used, or **cwm** (pronounced koom) in a Welsh context.

Cirques are invariably present above the sides, and at the heads of glacial troughs, or even close to sea level in some glaciated coastal mountains. Cirques are an extremely varied landform in terms of both size and shape. They may be as little as a few hundred metres wide (Fig. 3.13), yet the largest (the Walcott Cirque in Victoria Land, Antarctica) is 16km wide and has a headwall 3km high (Flint 1971: 133). Cirques in the steep terrain of alpine regions tend to slope outwards and to be poorly developed. In areas of less pronounced relief they commonly contain lakes (tarns), filling rock basins or dammed by moraines. However, the length:height ratios from the lip of a mature cirque to the top of the headwall are remarkably constant, ranging from 2.8:1 to 3.2:1. Thus, for the Western Cwm of Everest it is 3.2, for

Figure 3.13 (a) A well formed small cirque, Addacomb Hole, in the English Lake District. Formerly occupied by a moraine-dammed tarn, the cirque is now dry; the gullies cutting the moraine are clearly visible in the photograph. (b) Composite cirque named Coire an Lochain Uaine, below the summit of Cairn Toul, Cairngroms, Scotland. The overdeepened rock basin is occupied by a tarn.

the Blea Water Corrie in the English Lake District, a fraction of its size, it is 2.8 (Manley 1959).

Following analysis of the form of many Scottish cirques, Haynes (1968) found that their longitudinal profiles can best be described by logarithmic curves of the form:

$$y = k(1-x)^{e-x}$$

where x is the distance from headwall to lip, y is the depth of the cirque from the headwall to the basin, and k is a constant. The k values reflect how well the basin is developed. At the ends of the range of k values, $k = 2$ is characteristic of a deep cirque with a tarn and a steep headwall, whereas $k = 0.5$ is typical of a relatively open cirque with a gently inclined headwall (Fig. 3.13a).

The detailed form of cirques is closely controlled by the structure of the rock, in particular jointing and bedding. A complicating feature, however, is that many cirques are composite in nature, with small cirques formed at a later, less intensive stage, of glacierization, within the major feature. Many British mountain areas have cirque-within-cirque forms. A good example is Coire Bà, one of the largest in Britain, cut into the east face of the Black Mount in the Grampian Highlands. Here, several small cirques are incised into the headwall of the main cirque. In the Cairngorms, Coire an Lochain Uaine is a double cirque that forms part of a whole complex of cirques around the headwaters of the River Dee (Fig. 3.13b).

Some cirques have a composite long-profile, representing a number of cirques at different elevations, known as cirque stairways. One of the best examples in Britain is on Snowdon. Here, the large cirque of Cwm Llydaw with its over-deepened basin containing a tarn, has a precipitous headwall, breached part way up by a smaller cirque, Cwm Glaslyn, also with a tarn. Above the latter, there is a small incipient cirque directly below the summit of Y Wyddfa. The stream connecting them falls over steep cliffs, the headwalls of the successive cirques. In full glacial times ice formed a continuous stream with icefalls. At less intense phases, the higher cirque may have supplied the lower ones with avalanche debris, so for a time at least two of them may have contained separate glaciers, but equally likely is that each represents a different level of glacierization. Another possibility is that accumulation may have been controlled by the wind so that the greatest accumulation was at the foot of headwalls irrespective of altitude.

One of the few comprehensive studies of a cirque glacier is described in a classic monograph edited by Lewis (1960). Lewis' team examined the relatively simple Vesl-Skautbreen in the mountains of Jotunheimen in Norway. A tunnel was hand-dug through the ice, by a team of volunteers, to the cirque headwall in two places and a comprehensive three-dimensional picture of the velocity distribution and structure was obtained. The dominant structure is annual stratification which, as it moves down through the glacier, changes from a surface-parallel down-glacier dip to an up-glacier dip near the snout. Both this and the velocity measurements demonstrate that ice-flow has a rotational slip component, thereby enabling the greatest erosion to occur directly below the level of the equilibrium line. This rotational component of flow is thus held to be responsible for the over-deepening that ultimately creates a basin for the tarn.

The initiation of a cirque is a matter for debate, but accumulation of snow on leeward slopes, especially in depressions (nivation hollows), is likely. Once successive years of snowfall have accumulated and turned to ice, flow takes place, and downward and backward erosion commences. As the floor forms, it is abraded, while dilation of the bedrock and freeze–thaw, resulting in rock fracturing and block removal, takes place at the headwall.

Roches moutonnées

Features similar in size to whaleback forms but which are only streamlined on the side facing up-glacier are known as **roches moutonnées**, a term introduced in 1787 by De Saussure because of their similarity to *moutonnées*, the wavy wigs that were in fashion at the time (Embleton & King 1975: 152). They are much more common than whaleback forms and have an asymmetrical form as a result of glacial plucking along joints on their downstream side; they are particularly well developed in jointed crystalline rocks. Rock fracturing, induced by the impact of ice-embedded stones, is probably also important on the crest of a roche moutonnée.

Large areas may be covered by roches moutonnées; they provide one of the most useful criteria for determining the former directions of ice-flow. A very striking characteristic of glacially eroded terrain is the contrast between the smooth appearance of the eroded surface looking in the direction of former ice-flow and the craggy appearance looking up stream. In many areas, the form and size of roches moutonnées are related to the other structures in the rocks besides joints, e.g. foliation, faults, dykes and alternating hard and soft lithologies.

The development of roches moutonnées is probably related to pre-existing hillocks. The suggestion by Carol (1947), following observations beneath the Oberer Grindelwaldgletscher

Figure 3.14 Large (250 m high) roche moutonnée in granite, Lembert Dome, Yosemite National Park, California. The smooth, abraded upstream face to the right contrasts with the steep, plucked face to the right.

in Switzerland, that plucking was initiated by meltwater freezing in joints in a zone of lower pressure, beyond the crest of a protrusion in a glacier bed, is supported by observations beneath various other temperate glaciers and theoretical studies of basal sliding. Roche moutonnée-like forms, thought to be the result of the same processes, often occur as partially streamlined bosses on steep hillsides.

Most roches moutonnées range from a few tens to a few hundreds of metres in length, and several to tens of metres high. There are also smaller and very much larger ones in some areas. In parts of Scandinavia basement rock hills some hundreds of metres high are commonplace in the more subdued terrain. A well known large roche moutonnée is the Lembert Dome in the Yosemite National Park; it rises about 250 m above the Tuolumne Meadows (Fig. 3.14).

Low-lying areas that have been heavily scoured, such as the Baltic and Canadian shields, and the North-West Highlands of Scotland display many, closely spaced roches moutonnées as well as small rock basins. Many of these do not have a consistent shape because of complex structural controls, such as variably orientated faults and foliations.

In the geological record good examples of roches moutonnées can occasionally be found. Abraded bedrock surfaces in the Neoproterozoic sequence of Mauritania display a number of well developed roches moutonnées, and others may be found beneath the Late Palaeozoic Dwyka Tillite of South Africa.

Crag-and-tail features

Crag-and-tail features (also called **lee-side cones**) are composite forms, being the result of both erosion (the crag) and deposition (the tail, comprising lee-side till). The up-glacier craggy end is often steep and rough, while the tail is smooth and reminiscent of a drumlin. Britain's best known feature is Castle Rock in Edinburgh.

3.3.4 Rock crushing

Many striated and polished rock surfaces also show a variety of other small-scale features, normally a few centimetres in plan (but occasionally up to 2 m) and of crescentic appearance. These features have commonly been referred to collectively as friction cracks, but this is a poor term since Drewry (1986), among others, has demonstrated that they are the result of rock crushing, under the repeated impact of debris in basal ice on a segment of bedrock, until failure occurs.

There are several types of crush feature, including **lunate fractures, crescentic gouges, crescentic fractures** and **chattermarks** (Fig. 3.15) . They tend to lie with the concavity pointing either in or against the direction of flow. The principal fractures in these features tend to dip in a down-glacier direction. However, some authors have argued that this is an unreliable criterion for determining ice-movement directions, although crescentic gouges do seem to be reliable. As with striations, care must be exercised in using crescentic fractures as precise indicators of glacier flow. These features, notably chattermarks, are also good indicators of the jerky nature of glacier flow, and the associated stick–slip behaviour.

3.3.5 Erosion by ice combined with frost action

In mountainous terrain, valleys and cirques may extend progressively backwards as a result of erosion by ice. Frost-shattering is also an important process, and glaciers facilitate the removal of the debris. Progressive destruction of a mountain mass in this manner generates a family of landforms, which are essentially relict features. Two cirques that are being enlarged and approaching each other may eventually cut through the intervening ridge to produce a narrow, serrated ridge or **arête**, breached in places by cols. Three or more cirques eroding backwards against a single mountain mass may eventually meet and leave behind a pyramidal peak or **horn** (Fig. 3.1). Both horns and arêtes are common in areas that have undergone prolonged or active glacial erosion. If an ice sheet competely inundates these features at a later stage, then they may be worn down into dome-like peaks and rounded ridges.

Arêtes

The term arête is French, and is applied to the knife-edge ridges that are so abundant in the Alps. Arêtes are mainly the result of backward erosion by adjacent cirque or valley glaciers, during which the rock dilates and fractures along joints. Frost-shattering facilitates the development of a jagged ridge, on which upstanding pinnacles are referred to by climbers as **gendarmes** ("policemen").

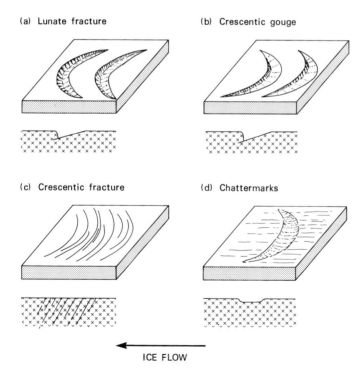

Figure 3.15 Small-scale features of glacial erosion with cross sections, formed largely as a result of rock crushing (based mainly on Embleton & King 1968: Fig. 6.3).

Figure 3.16 Narrow ice-draped arêtes on Mount Tasman, Southern Alps, New Zealand.

Figure 3.17 Curving arête leading from Carn Mòr Dearg towards the summit of Ben Nevis (off picture to the right) in Scotland. The range of hills in the background, the Mamores, also comprises arêtes, together with cirques.

Arêtes are widespread in alpine terrains all over the world. The Western Cordillera in Canada and Alaska as well as the European and New Zealand Alps, the Andes and Himalaya all have classic landforms (Fig. 3.16). The western parts of Britain and Norway have well developed arêtes arising from late glacial cirque activity despite being under an ice sheet for much of the Pleistocene epoch. Those well known to British fell walkers for providing sporting routes to the summits include the ridge of Crib Goch on Snowdon (Wales), Striding Edge on Helvellyn in the Lake District, the Carn Mòr Dearg ridge on Ben Nevis (Fig. 3.17) and the Cuillin Ridge on the Isle of Skye (Scotland).

Cols

Mention has already been made of one special type of **col** – the watershed breach which is the product of abrasion as ice escapes from its confining trough. Many cols, however, are simply the result of backward erosion of adjacent cirque or valley glaciers by rock fracturing, combined with lowering locally of the intervening arête (Fig. 2.31). A col may thus be a low notch, approachable only by ascending the steep headwalls of the adjacent cirques. The arête leading down to the col may similarly be precipitous. Many Alpine passes are of this nature. In Britain erosion has only proceeded to this degree in relatively few places. The Cuillin Hills on Skye provide the best examples.

Horns

Backward erosion by three or more cirque glaciers in combination with frost-shattering, may lead to the intersection of the cirque headwalls, thus isolating an upstanding mass of rock, called a **horn**, a term loosely derived from the German word for a peak. A horn represents the stage at which all the original smooth highland has been eroded. The horn is epitomized by the Matterhorn (Fig. 3.1) which stands astride the Swiss–Italian border. Three nearly symmetrical ridges rise steeply towards a pyramidal summit, each face carrying a steep cirque glacier. Other well proportioned horns are Mount Assinboine in the Canadian Rockies, Mount Aspiring in New Zealand and K2 in the Karakorum Range in Pakistan. Most horns, however, lack this symmetry owing to the uneven backward erosion of cirques. Some horns even have four or more arêtes leading up to their summits. Few British peaks acquire the status of a horn, but examples are Schiehallion in the Grampian Highlands and Cir Mhòr on the Isle of Arran (Fig. 3.3b). The Cuillins on the Isle of Skye have steeper sides but less well developed individual forms.

Nunataks

Rock outcrops ranging from less than a kilometre to hundreds of kilometres across that are surrounded by ice are known as **nunataks**, a term derived from the Inuit language. They include the last relics of mountains that have been subjected to valley-glacier, cirque-glacier or ice-sheet erosion, as well as intensive frost-shattering. At the other extreme, nunataks may be represented by entire mountain ranges that have only been affected by erosion on their flanks, and the intervening surface, although heavily frost-shattered, may preserve its preglacial form. Both end members and everything in between exist in Antarctica, the most extensive development being represented by the Transantarctic Mountains which stretch across the entire continent. The Greenland ice sheet has many nunataks towards its periphery, where it begins to spill through the coastal mountains, and nunataks are common in areas that have highland ice fields such as Ellesmere Island and Spitsbergen. Smaller nunataks are also characteristic of heavily glacierized alpine regions.

3.4 Structural geological controls on glacial erosion

The processes of glacial abrasion, rock fracture and block removal are facilitated if there are weaknesses in the bedrock. Major structures, such as faults, commonly provide the main discharge routes from ice caps and ice sheets. However, the fault-controlled valleys may well have been in existence prior to glaciation and have been floored by deeply weathered material that subsequently was easily picked up by the ice. As already described, one of the Antarctic's major discharge routes, the Lambert Glacier, flows in a deep graben, bounded by parallel faults. Many fjords in the Canadian Arctic also follow graben structures. Ice streams in the Scottish and Scandinavian ice sheets flowed along faults. Distinct preferred trends of lochs and glens are visible in satellite images of northwest Scotland; they follow a set of NW–SE trending faults, as well as a major structure, the Great Glen Fault, which crosses the country from NE–SW. Another good example is in East Greenland where two sets of intersecting faults, one running east-west, the other NW–SE, are now followed by fjords, isolating individual fault blocks.

On the other hand, some glaciated valleys may not show any clear relationship to bedrock structure. The English Lake District is a case in point: the radial drainage system here was developed on a dome of Late Palaeozoic and younger rocks that developed in early Tertiary time. Ice continued the down-cutting below an unconformity into folded Early Palaeozoic rocks, without adapting to the differently orientated structures in the lower strata. This phenomenon is known as **superimposed drainage**.

Landscapes of areal scouring display many close links between landform and bedrock structure. Complex assemblages of ribbon lakes, small straight valleys, gullies and roches moutonnées all show a close relationship to bedrock foliation and cross-cutting faults, and the land surface may give the appearance of being strongly dissected. The Early Proterozoic and Archaean rocks in the Lewisian basement of North-West Scotland (Fig. 3.18) and matching rocks in the Laurentian Shield of Canada often contain ribbon lakes and gullies following these structures.

Cirques are not necessarily located in areas of structural weakness, but their detailed form is often controlled by bedding foliation or jointing. Backwalls tend to be steep if there is a set of exploitable, near-vertical joints. Similarly, the detailed form of roches moutonnées reflects the same sort of structural feature.

Figure 3.18   The Loch Laxford area, Sutherland, Scotland, where the early Precambrian Lewisian Gneiss has been subjected to "areal scouring" by an ice sheet. The strong structural control is determined by faults, foliation and dykes in the gneiss. (With permission of Cambridge University Committee for Aerial Photograph; photograph no. NC 186503, copyright reserved).

3.5 Preservation potential of erosional forms

Most erosional landforms are known from land areas, and one might expect that their preservation potential is low, as erosion often continues after the ice has receded. However, as a glacier recedes it commonly deposits a layer of till on scoured bedrock, which protects it from subaerial weathering. Thus, many fresh Pleistocene and older eroded surfaces have been found where the overlying till or tillite has been removed.

On a longer timescale, erosional forms have less chance of survival, unless they are rapidly buried beneath a thick sedimentary pile in a subsiding basin. Nevertheless, there are a surprising number of striated pavements and roches moutonnées in the geological record, examples having been mentioned above, but larger scale features, such as glaciated valleys, cirques or horns, have only rarely been recognized.

4 Environments of terrestrial glacial deposition

4.1 Introduction

This chapter concentrates mainly on the environments where till and associated sediments are deposited, which are more pronounced in the lower part of a glacier system. It has been said that till is more variable than any other sediment known by a single name; partly because of this, there is, as yet, little agreement as to a suitable definition (see Ch. 1). In this section the various characteristics of tills and related terrestrial deposits, together with the landforms resulting from deposition, are considered.

4.2 Processes of terrestrial glacial deposition

Until comparatively recently, glacial geologists interpreted the mode of deposition of tills on the basis of their examination of Pleistocene deposits, and rarely considered the processes actually occurring in present-day ice sheets, with the result that many misconceptions arose. The position has altered dramatically over the past 20 years or so, following many detailed studies of the processes involved, notably by Boulton in the 1960s and 1970s (reviewed by Sugden & John 1976), and others subsequently (reviewed by Dreimanis 1989). A comprehensive account of glacial depositional processes is given by Sugden & John (1976: Ch. 11). Although modern studies have drawn attention to the complexity of glacial environments, the better understanding obtained of subglacial processes allows us to make meaningful interpretations of Pleistocene and pre-Pleistocene glacigenic sequences. The relation of the most common types to a receding, wet-based glacier is shown in Figure 4.1.

4.2.1 Subglacial lodgement

The zone of erosion beneath a typical glacier is followed downwards by a transition to a depositional regime beneath the actively moving ice, usually as the glacier is losing its erosive capacity (Fig. 4.1). Pressure-melting allows material to be released from the dirty basal

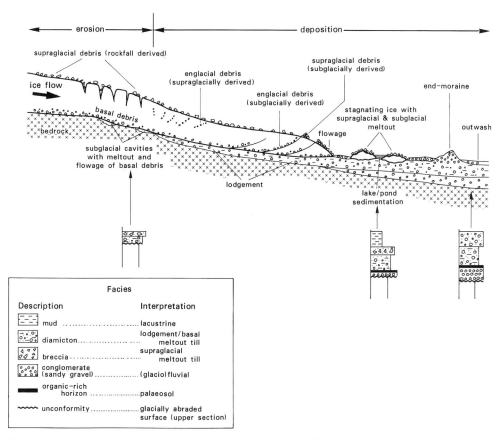

Figure 4.1 Cross-sectional view of a typical temperate glacier to illustrate the various types of glacigenic sediment deposited as the ice recedes.

ice, particularly on the upstream side of bumps. Together with basal shear within the ice/debris mixture, which facilitates renewal of the debris supply, this process leads to debris being plastered on to the bed, and progressively built up. This process, known as **lodgement**, occurs both on bedrock and older till surfaces, and leads to the filling of bedrock irregularities, so smoothing out the glacier bed. Some till may be squeezed into cavities down stream of a bump. Lodgement occurs beneath both advancing and receding glaciers, and the details are now well established from studies by Boulton (1970, 1971) on glaciers in Spitsbergen. A common characteristic of lodgement till is the presence of shear structures formed as the ice overrides the unconsolidated deposit. Where preserved in older deposits, the pattern of such structures is potentially useful as a palaeo-ice-flow indicator. Shear stress also results in the orientation of stones in the till developing at an angle of 45° to the bed (parallel to the plane of maximum shear stress), but, as the material continues to deform, stones rotate so that they tend to approach parallelism with the bed, thus providing an additional means of establishing the direction of flow in old lodgement tills.

Deposition of lodgement till may be strongly affected by the transfer of debris-rich basal ice into an englacial or even supraglacial position by movement along shear or thrust zones

111

Figure 4.2 Deposition of subglacial melt-out till and reworking by meltwater at the snout of the Vadrec da Morteratsch, Switzerland. Note the predominantly subrounded and subangular character of the clasts.

within a glacier undergoing compressive flow (§2.10). Strictly speaking, the subsequent release of such deposits is by the melt-out process.

The rate of subglacial till deposition can be considerable, and 6 m per century is regarded as reasonable (Sugden & John 1976: 218). Contrary to the belief of glaciologists up to about 20 years ago, it is now known that much of the Antarctic and Greenland ice sheets are at the pressure melting point at their base. Thus, the deposition of thick tills and tillites on a continental scale as recorded in Pleistocene and earlier successions is not difficult to explain.

4.2.2 Melt-out and flowage

Melting out of debris can occur either subglacially or supraglacially, a process that is particularly active at the margins of glaciers and ice sheets, where the ice is practically stagnant (Fig. 4.2). Geothermal heat is largely responsible for the deposition of subglacial melt-out tills. If the water is able to escape without disturbing the sediment, the resulting till may preserve traces of the original relationship of debris to the ice structures. Melt-out till is commonly deposited in unstable situations, and is thus prone to flowage.

During the past 50 years, frequent reference has been made to the possibility of flowage of saturated till at the base of a glacier. For example, Gripp (1929) suggested that some englacial bands were the result of till being squeezed up into basal crevasses, a process Mickelson (1971) has observed in action. Hoppe (1957) has identified certain types of hummocky moraines resulting from same process. More recently, Boulton (1971, 1972) has noted that flowage can occur where thick unfrozen subglacial till exists beneath temperate ice. Basal crevasses are not particularly common beneath glaciers, as it happens, unless the ice is at an advanced stage of decay, but there are other spaces into which till can be squeezed, in particular hollows down stream of obstacles, abandoned subglacial stream channels and moulins. Stones in wet lodgement till may undergo re-orientation if subsequently affected by flowage and the final fabric may be very different from till that has not been affected. Features of "plastic" deformation, particularly folds, may aid the recognition of such till.

Debris at the surface of a glacier may have a basal derivation, especially near the margins of cold glaciers and ice sheets (Fig. 2.35) or, in the case of valley glaciers, it may have accumulated as a result of subaerial weathering of rock faces. This debris melts out in summer as a result of ablation. A layer of debris generally retards melting of dirty ice in comparison with clean ice, so the resulting ice surface tends to acquire debris ridges. As the debris is lowered it tends to undergo sliding and re-orientation, and rarely reflects the original fabric in the ice after deposition as till. A thick layer of supraglacial debris tends to be unstable and, as with basal melt-out till, is often capable of flowage.

Regardless of whether the sediment is deposited supraglacially or subglacially, the process is particularly common near the snouts of receding or stagnating glaciers. The process has been particularly well studied by Boulton (1968, 1971) at the margins of various Spitsbergen glaciers and by Lawson (1981, 1982) at the snout of Matanuska Glacier in Alaska (Fig. 4.3a). The principal factors affecting debris flowage are the gradient of the bed on which it rests, whether it be ice or the glacier bed itself, the bed roughness, the amount of meltwater available for enabling the debris to become more fluid, and the fabric of the debris itself. Saturated till may flow on the gentlest of slopes. According to Boulton, three types of flow can be distinguished which, in decreasing order of rate of movement, are:

(a) mobile flow, in which stones may tend to settle towards the bottom producing a crude sorting pattern (Fig. 4.3b);

(b) semi-plastic flow, which often begins by slope failure along an arcuate slipface (Fig. 4.3c) and is perceptible as a slow-moving lobate tongue (Fig. 4.3d) – boulders may sink, fold structures may form, and washing by meltwater may produce laminations;

(c) downslope creep, which involves less water and is not perceptible to the eye – although the till has a fabric that is subject to alteration and may acquire a weak foliation, it is unlikely to develop fold structures. Such till, in contrast to other types of flow till, is compact and massive.

Strictly, a "flow till" is resedimented and some authors (e.g. Lawson 1979) have argued that the term should be abandoned. The often too subtle differences between such sediments and "non-disaggregated" tills arguably makes such a step premature.

(a)

Figure 4.3 Glacigenic sediment flowage in close proximity to Alaskan glaciers. (a) General view of the snout of the Matanuska Glacier, illustrating a proglacial area that has been almost totally disturbed by flow processes. (b) The debris-covered snout of Matanuska Glacier illustrating mobile flow in the foreground; a slurry of fine material to the bottom of which coarser material settles. (c) "Semi-plastic flow" often begins with slope failure within till as here, a few hundred metres from the snout of Matanuska Glacier. (d) Semi-plastic flows commonly terminate in a lobate tongue; here, failure of a till-covered slope has occurred, the debris flowing out across the surface of Orange Glacier.

(b)

(d)

(c)

4.2.3 Sublimation of debris-rich ice

Sublimation is the process of ablation whereby ice is vaporized directly without passing through the intervening liquid phase. Some authors have claimed it to be of negligible importance, but in the cold, arid parts of Antarctica, such as the Dry Valleys of Victoria Land, where temperatures rarely exceed 0°C, it is a common process (Shaw 1977). Here, debris-rich ice ablates from the surface downwards, producing a loose sublimation till which inherits the foliated structure from the ice (Fig. 4.4). Outside Antarctica the process is rare, however.

Figure 4.4 Formation of sublimation till from debris-rich basal ice, Taylor Glacier, Victoria Land, Antarctica. The darker material is the dirty ice, and the lighter coloured cap is crumbly sublimation till that is inheriting the foliated structure from the ice.

4.2.4 Ice-push

During the winter months, when ablation ceases, it is common for even a generally receding glacier to advance a short distance and push up a small ridge of till and fluvioglacial material (Fig. 4.5). To produce a series of annual push moraines (e.g. Worsley 1974). More spectacular push-moraines often occur at the snouts of advancing glaciers, but they are the product of glaciotectonic deformation (§4.2.6).

4.2.5 Subglacial chemical processes

The importance of chemical processes involving calcium carbonate beneath terrestrial glaciers was first noted by Ford et al. (1970). Other chemical processes involving silica have also been identified. It is now known that comminution of carbonate rock particles creates a rock flour that is highly reactive chemically. Carbonate is thus readily dissolved and precipitated (Fairchild 1983, Souchez & Lemmens 1985). Isotopic studies beneath Swiss glaciers in limestone areas have revealed that chemical action can be important at the ice/bedrock interface. Souchez & Lemmens (1985) and Sharp et al. (1989) found that calcite was formed by precipitation, as patchy coatings and cornices up to a few centimetres thick, on polished and striated

Figure 4.5 Small (2 m high) push moraine from the preceding winter's advance at Steingletscher, Switzerland. Note the person on the ice-abraded bedrock with p-forms in the background.

Figure 4.6 Chemically precipitated calcite coatings, with striations and cornices, on limestone bedrock, Glacier de Tsanfleuron, Switzerland. Ice-flow was from left to right.

bedrock. Some of the coatings themselves were found to be striated (Fig. 4.6). Also, calcite-coated pebbles and thin limestone crusts were present in recently deposited basal tills.

Calcite is released when meltwater refreezes. Partial freezing is accompanied by concentration of salts in the residual water, eventually resulting in saturation. During freezing, water in equilibrium with the growing ice is progressively impoverished in heavy isotopes, notably oxygen-18. Sliding over protuberances on the bed results in pressure-melting on the stoss side and refreezing on the lee side, thus leading to carbonate deposition. Fixed bedrock protuberances give fluted or furrowed coatings on the lee side. The ^{18}O isotopic composition indicates that the origin of the initial water which gives the precipitate is the basal ice layer, not true glacier ice.

Calcium carbonate coatings on pebbles and calcite layers in Neoproterozoic tillites have been reported from China (Wang Yuelun et al. 1981), the Sahara (Deynoux 1980) and Svalbard (Hambrey 1982). Such carbonate coatings may form in a variety of settings, including subglacially, proglacially and pedogenically; or arise later during burial diagenesis or metamorphism (Fairchild 1993, Fairchild et al. 1993).

4.2.6 Glaciotectonic deformation

Glaciotectonic deformation is now accepted by many researchers to be an important process, but until the publication of the book by Aber et al. (1989), this was not reflected in the general literature. The possibility that glaciers could deform shallow crustal rocks and sediments was recognized well over a century ago, by Lyell during studies in Norfolk, England and the Italian Alps. Glaciotectonic processes can be recognized throughout glaciated terrains where sedimentary bedrock and thick drift deposits occur. Strictly speaking, glaciotectonic processes involve erosion as much as deposition; their discussion in this chapter reflects the fact that they are most evident in nearly contemporaneous sedimentary sequences. The application of structural geological principles to the study of glaciotectonism has revolutionized our under-standing of the resulting products.

Glaciotectonic deformation takes place in a variety of settings – in front of a glacier, beneath the ice margin or under the middle part of a thick ice sheet. Glaciotectonic deformation operates in any topographic setting, both during advancing and recessional phases, and involves all manner of sediment types in frozen, saturated and dry states.

The principal factors that are important for the genesis of glaciotectonic phenomena are as follows (from Aber et al. Table 9.1):
– lateral pressure gradient;
– elevated groundwater pressure;
– ice-advance over permafrost;
– ice-advance against a topographic obstacle;
– lithological boundaries in the substrate;
– surging of ice lobes;
– subglacial meltwater erosion;
– damming of proglacial lakes;
– thrusting in front of the ice;
– compressive flow with basal drag;
– shearing of fault blocks up into the ice.

Glaciotectonic deformation takes place when the stress transferred from the glacier exceeds the strength of the material beneath or in front. The material may be subject to both brittle and ductile deformation, typical structures being thrusts and folds, respectively. Sometimes, the glacier detaches a sediment-mass or slab of bedrock along a plane of décollement, incor-porating it into the body of the ice before depositing it (cf. §2.10.3; Fig. 2.36). The structure of the glaciotectonically deformed sediment is strikingly similar to that of a mountain belt, such as the Alps or the Rockies, but on a much smaller scale.

4.3 Terrestrial glacigenic facies and facies associations

4.3.1 Boundary relations and geometry of glacigenic units

Most till is deposited during glacial recession after the ice mass has already eroded bedrock or the underlying unlithified sediments. Thus, tills and tillites commonly unconformably overlie striated bedrock surfaces, or more rarely boulder pavements. These surfaces are irregular, but the basally deposited till tends to fill the hollows and smooth out the relief. The top of a basally deposited till may be quite regular and in distinct contrast with supraglacial till lowered on to it. The upper surface of the latter may, however, be extremely hummocky, reflecting uneven down-wastage of the ice, and the whole picture may be confused by till flowage. Overlying sediments may be fluvial or aeolian, forming a distinct break and tending to smooth out the surface. Within the glacigenic facies are intimate associations with fluvial materials, which occur as lensoid or channel-like bodies. Till may be spread over many hundreds of square kilometres as a deposit varying in thickness from a few metres to tens (or exceptionally a few hundreds) of metres, as a result of accretion during successive advances. Entire sequences may become disrupted and structurally complex during glaciotectonic processes, and the final geometry will reflect large-scale movements of material *en masse*.

4.3.2 Sedimentary structures

Terrestrial tills characteristically lack well defined bedding or lamination. However, lodgement tills may have a distinct fissility or foliation, resulting from shearing as the glacier overrides the unconsolidated material. Exceptionally, melt-out tills have a weak "bedding" acquired from debris-rich dirty layers parallel to basal foliation in the ice. Flow tills, as already discussed, may be subjected to slight grading as the heavy stones settle during flowage and, in addition, bear evidence of slump folding with axial planes parallel to the depositional surface and fold axes normal to the flow direction. Differential loading of till on to other soft sediments may result in convolutions and injection phenomena, although these features are rare.

Till, of course, is rarely deposited in isolation from other sediments. Glaciofluvial deposits, for example, may be preserved as lenses or channels within a till unit and may have structures which are associated with normal braided river deposits, such as cross-bedding. Lacustrine sediments or aeolian sands may also form lensoid features within the till sequence.

4.3.3 Facies

The principal lithofacies in the terrestrial glacial environment are the products of direct glacial, fluvial, lacustrine and aeolian processes (see also Chs 5 & 6). Of the sediments released directly from the ice, diamicton is the principal lithofacies, especially in its massive form. Generally, it is interpreted as lodgement, melt-out or flow till, depending on the relationships with adjacent beds. Also common are breccias, derived from angular, supraglacially transported rockfall material. These facies may be intimately associated with conglomerates and sandstones of fluvial origin, laminated muds of lacustrine origin, and cross-bedded sands of aeolian origin.

In general, terrain of alpine character produces more supraglacial debris with predominantly angular clasts than more subdued terrain. On the floors of many alpine valleys, occupied by temperate glaciers, there is so much reworking by meltwater, that little till can survive, the bulk of the sediment being glaciofluvial. The thickest diamicton sequences occur in the Arctic, where cold glaciers carry a much greater basal debris load than their alpine counterparts.

4.3.4 Facies associations

Unlike many glaciomarine sedimentary records, terrestrial sequences tend to be very incomplete. If a wet-based glacier re-advances over sediment released earlier, recycling, removal and incorporation into the new deposits frequently occurs. Thus, not all successive advances are preserved, and the dominant facies tend to reflect the final recession of the ice. Complex multilayered diamict sequences relating to one glacial phase may arise from melt-out processes with or without flowage. At the base of a glacier, deposition of melt-out till may begin after lodgement has ceased, e.g. during glacial recession, while supraglacial debris on the glacier surface will be deposited on top of the basal facies.

It is also apparent that flowage almost certainly complicates the picture, and recognition of the depositional characteristics of a typical till sequence will require very close scrutiny, and even then it may not be possible to make a meaningful interpretation.

4.3.5 Facies architecture

Terrestrial glacigenic sequences are characterized by their complexity and rapid vertical and lateral facies changes. Although there are few places where extensive vertical sections of Quaternary age demonstrate the two-dimensional geometry of facies, let alone the three-dimensional architecture, we can demonstrate the expected disposition on the basis of our understanding of the depositional processes. However, better exposed pre-Pleistocene terrestrial glacigenic sequences may aid our assessment of facies architecture.

4.4 Physical, chemical and mineralogical characteristics

4.4.1 Grain-size distribution

Tills and tillites are characterized by extremely poor sorting, containing all size classes from clay to boulder (Fig. 4.7).

Grain-size distribution characteristics of till are dependent on whether it is supraglacially or basally derived. Triangular plots of the sand, silt and clay fractions have been presented for a variety of lodgement- and supraglacially derived tills from North America and Britain (Sladen & Wrigley 1983). Material at the base of a glacier is subjected to attrition and breakdown into clay-size particles, and so is finer grained than supraglacial material, but there is a degree of overlap (Fig. 4.8a).

The nature of the terrain also influences the grain-size distribution. Data from three types of terrain are depicted in Figure 4.8b. Shield terrains (1) and glaciated valley terrains (2) yield coarse, sandy, clast-rich tills, whereas glaciated sedimentary lowlands (3) yield tills that are rich in fines, especially where lacustrine and marine sediments have been incorporated into the tills.

The degree of glacial transport also influences grain-size distribution in tills. Most tills have a bimodal or polymodal distribution, the latter tending to reflect the greater distance of transport. The composition of the principal rock type making up the till also influences the grain-size distribution. Coarse-grained igneous and metamorphic rocks, such as granites and gneisses, and sandstones tend to give rise to sandy tills, whereas the matrix of fine-grained sedimentary rocks is dominantly clayey or silty.

(a)

(b)

Figure 4.7 Poorly sorted character of massive diamicton, interpreted as lodgement till. (a) Till deposited by the Devensian ice sheet, Loch Lurgainn, North West Highlands of Scotland. (b) Matrix of a lodgement tillite of Neoproterozoic age, Wilsonbreen Formation, Svalbard. The dark background is clay grade, most of the white grains are silt grade, and the larger fragments are of sand. The field of view horizontally is approximately 5 mm.

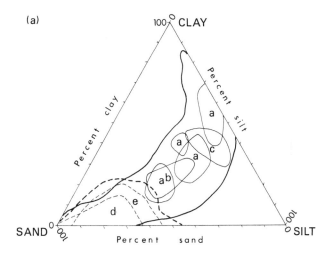

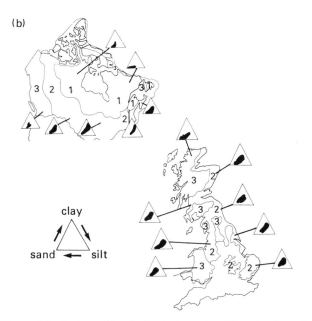

Figure 4.8 Grain-size distribution in the sand and finer fractions of tills from North America, Iceland and Britain. (a) Thin solid lines – envelopes for lodgement tills from A, Ohio; B, Northumberland; C, Ontario; thin dashed lines – envelopes for coarse-grained supraglacial diamicts from D, Iceland; E, Scotland. The heavy solid and dashed lines enclose all tills in these categories, respectively. (b) Grain-size distribution envelopes for lodgement tills from A, shield terrain; B, sedimentary lowlands of subglacial terrain; and C, glaciated valleys (from Sladen & Wrigley 1983, with kind permission of Pergamon Press, Oxford).

Figure 4.9 Basal till in a cold polar setting, Schirmacher Oasis, Dronning Maud Land, Antarctica. This till comprises high-grade metamorphic clasts in a sandy matrix; there is little silt and finer material.

Grain-size distribution is also influenced by the thermal regime of the ice mass, since the availability of meltwater facilitates the grinding down of the coarser particles. In Antarctica today, where the supply of meltwater is limited, tills tend to be sandy, in contrast with those associated with Northern Hemisphere glaciers where meltwater is abundant (Fig. 4.9).

4.4.2 Shape and surface features of clasts

A characteristic of till, unusual in other types of sedimentary rock, is that it contains stones of both local and far-distant (exotic) origin, usually well mixed and apparently homogeneous over wide areas. Many studies have been made on the provenance of stones in till but, as yet, few analyses have been made of pre-Pleistocene tillites. For Quaternary deposits, studies of boulder trains of a particular lithology in some cases have the potential for tracing ore bodies, although this method of prospecting is still in its early stages of development (Shilts 1976). The picture of apparent homogeneity of till is more complex than casual observation suggests. A statistical analysis of the Catfish Creek Till in Ontario (May & Dreimanis 1976) found that the matrix of a till is inherently more homogeneous than the stones, and that the lowest part of a till unit, containing locally derived material, is in fact so variable that it should be excluded when characterizing the entire unit.

The shape of stones in till varies from rounded to angular, but normally subangular to subrounded varieties predominate. Debris falling on to a glacier surface generally is shattered and angular, and subsequently is transported passively in a supraglacial or englacial position (Fig. 4.10a). Basal debris is subjected to abrasion and rotation, so that rounding of the irregularities takes place, and the stones have a broader range of shapes with subangular and subrounded being dominant (Fig. 4.10b). These characteristics are inherited in lodgement till (Fig. 4.10c). Reworking by subglacial meltwater leads to further rounding (Fig. 4.10d). Boulton (1978) has shown that roundness combined with sphericity permits one to determine the means of transport and deposition. Three distinct sphericity/roundness fields can be recognized for supraglacially derived boulders in high-level transport (i.e. those derived from rock-fall), boulders from the zone of traction and boulders embedded in lodgement till (Fig. 1.4).

Apart from roundness and sphericity, stones in basal till often have specific shapes which indicate their glacial origin, in particular faceted, pentagonal and flat-iron forms (e.g. Wentworth 1936, Flint 1971, Boulton 1978). Many stones are clearly more pointed at one end than the other. Lithology does not appear to play an important part in determining the roundness of stones, since a study by Dowdeswell et al. (1985) of carbonate and crystalline clasts in a Spitsbergen tillite has shown no detectable difference in the shapes of the different

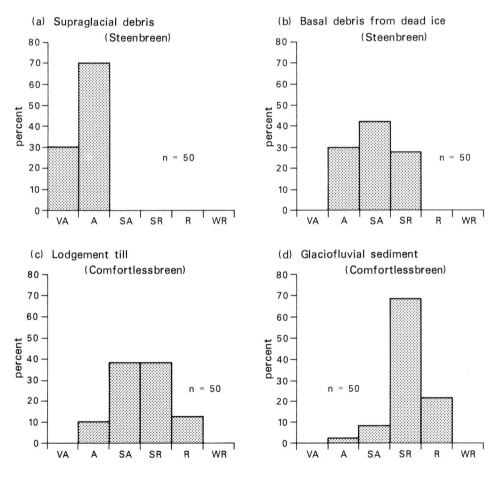

Figure 4.10 Range of shapes according to Power's roundness categories, in a variety of proglacial settings, Engelskbukta, Svalbard. (a) Supraglacial debris, Edithbreen; (b) debris from buried stagnant basal ice, Steenbreen proglacial area; (c) lodgement till from fluted ridge, Comfortlessbreen; (d) glaciofluvial material, Comfortlessbreen outwash area.

lithological populations. However, there are variations in Zingg shapes: bedded sedimentary or well foliated metamorphic rocks tend to produce blade, plate and flat-iron forms, partly because they tend to slide during traction, whereas gneisses and igneous rocks tend to be more rounded, as they are subject to rolling at the base of the glacier. Boulton (1978) and Sharp (1982) have further noted that the larger boulders (>0.5 m) which are deeply embedded in lodgement till have a characteristic form: a smooth, bullet-nosed up-glacier end, and a truncated down-glacier end (Fig. 4.11). In contrast, boulders loosely embedded in, or resting on, the till surface have no such clearly defined form, and although they may be faceted, they lack the streamlined form of embedded boulders. Tills often contain a proportion of angular stones, such as shales, which would not normally survive extended transport in other media. Supraglacial or englacial transport enables such "fragile stones" to be preserved.

Striations are commonly regarded as a distinctive feature of till stones (Fig. 4.12). However, supraglacially or englacially transported debris will have none, while perhaps only a relatively small proportion of basally transported stones will have striations. The proportion of stones with striations varies greatly (Dowdeswell et al. 1985), but generally it is of the order of 5–20%. Striations may occur on any part of a stone, but are more common on facets or on the surface of a stone projecting above lodgement till that has subsequently been overridden (Sharp 1982). Striations may be orientated completely at random, cross-cutting one another, such as when the stone has been subject to rolling, or may form subparallel sets, as for example, when little rotation has taken place (Fig. 4.12). The most regular striations occur on boulders embedded in the till; these form a single set, diverging slightly down-glacier (Boulton 1978) or they have a consistent orientation (Sharp 1982). Lithology has a marked influence on whether a stone is striated or not (§1.4.4).

Figure 4.11 Bullet-nosed boulder embedded in lodgement till near the now-stagnant margin of Burroughs Glacier, Glacier Bay, Alaska. Ice movement was from left to right.

Figure 4.12 Striated clast from the proglacial area of Aavatsmarkbreen, Spitsbergen. A single prominent set of parallel striae is cut by striae of many orientations. To the right is a prominent crescentic gouge indicating ice movement to the right. The stone is about 30 cm long.

In addition to facets and striations, some stones may bear other small-scale features of glacial abrasion of the types that are often found on abraded bedrock surfaces, such as crescentic fractures and chattermarks (Fig. 4.12). A feature, previously mentioned (§4.2.4), in areas of carbonate bedrock, is the presence of thin egg-shell-like muddy or calcareous coatings on some stones in tillites. The Chinese Sinian tillites have good examples of these, and there it has been suggested that, during glacial sedimentation, clay minerals were deposited, and dissolved calcareous matter was precipitated. Transport of till stones by meltwater soon results in rounding and the removal of the glacial markings; rolling on a stream bed may give rise to percussion marks instead.

4.4.3 Clast fabric

It has been known since the work of Miller in 1850 (Flint 1971) that stones in till often have a preferred orientation, and that this is parallel to the direction of ice movement (where known from other evidence). However, only in the past 30 years or so, has this property been used, to any extent, in palaeoenvironmental reconstructions. In most tills and tillites a preferred orientation can only be determined by detailed measurements, but occasionally is it visible to the naked eye. On the other hand, in some situations, till fabric measurements reveal stones orientated transverse to the direction of ice movement.

Fabric studies are useful for determining the direction of glacier flow during deposition, possible with greater accuracy than from features resulting from abrasion, especially if the stones are not equidimensional. The shape of stones, in particular roundness, sometimes has a marked effect on their orientation. For example, in lodgement till the long axes of the well rounded shorter stones tend to be aligned parallel to the flow direction, whereas longer less rounded and wedge-shaped stones are transverse to flow (Embleton & King 1968: 308). Thus,

127

because of the fact that both parallel and transverse fabrics can arise during glacial deposition, other sedimentological evidence must be examined before assessing the significance of a particular fabric. Sliding, shearing and postdepositional flowage can all determine the final fabric of a till. In general terms, we may expect the fabric of till to reflect both the position of debris in the ice during transport and the mode of deposition.

Figure 1.5 illustrates on stereographic projections (cf. Phillips 1971) the three-dimensional orientation of clasts in a variety of situations. Typically, clasts in basal ice show a strong preferred orientation. For example, Lawson (1979) at the snout of the Matanuska Glacier in Alaska found that in the debris-rich foliated ice from which basal melt-out tills were derived, the stone orientations plot as a single strong maximum, generally reflecting the local direction of flow, and lying in the plane of the foliation which itself was parallel to the glacier bed (Fig. 1.5, row A).

The basal melt-out till derived from the Alaskan glacier retained more-or-less the same fabric characteristics as the source ice (Fig. 1.5, row B) although if the pebbles were inclined in the ice, their angles of repose decreased on deposition. Supraglacial melt-out till also preserved the original fabrics of the ice except there was some dispersion of pebbles. Modification of the supraglacial till is often pronounced after deposition.

In lodgement till, the preferred orientation of stones tends to be parallel to the direction of local glacier movement, as one would expect. In addition, it is common for stones to dip up-glacier, rather in the manner of pebble imbrication in stream beds. Lodgement tills tend to have rather broad, even girdle-like maxima fabrics on a sterographic projection (Lawson 1979, Dowdeswell et al. 1985; (Fig. 1.5, row C). On the other hand, under certain stress conditions, such as strong compressive flow, stones may become orientated transverse to flow (Dreimanis 1976). Boulton (1971) found similar strong fabrics in Spitsbergen, but with transverse-to-flow orientation fabrics in zones of compressive ice-flow and parallel fabrics in zones of extending flow.

In the case of Matanuska Glacier, flowage of till is the dominant process, to the extent that only 5% of the till has been unaffected. These flow tills (or sediment flows as Lawson preferred to call them) are varied in character and have complex fabrics ranging from fairly strong alignment parallel to flow where water content was high, to weak multi-maxima fabrics as the water content decreased, although this pattern is not always consistent (Fig. 1.5, row D).

All these fabrics contrast markedly with the pattern resulting from the settling of basal glacial debris through a water columns, which tends to show a random distribution within the horizontal plane (Fig. 1.5, row E).

Most fabric studies have been undertaken using pebbles and cobbles, but it is also possible to make use of elongate grains of sand size under the binocular microscope. This technique only permits us to derive two-dimensional data, but nevertheless is especially useful for studies of core material (see Ch. 8). Comparisons between the orientations of sand grains, pebbles and cobbles, and boulders in a lodgement till in Switzerland demonstrate that reproducible fabric results are obtainable regardless of grain size.

4.4.4 Mineral composition of the matrix

Mineralogical analyses of terrestrial tills and tillites are relatively unusual. However, the technique has proved especially useful for tracing economic mineral deposits. This can be done, for example, by establishing palaeo-ice-flow directions from the distribution of heavy minerals (e.g. garnets) in the matrix or in ore-bearing boulders. Glaciers appear to disperse material so that the concentrations of minerals, elements or rock types reach a peak close to the source with an exponential decline in the direction of transport (Shilts 1976). The detection of such dispersal tails is a key feature in mineral exploration in till-covered areas. An interesting example of successful mineral exploration with limited resources is demonstrated by the Geological Survey and mining companies in Finland. In that country the populace have been encouraged to send by post, free-of-charge, any rock of exceptional appearance, with sizeable rewards for good samples. From 10000 samples mailed each year, 5% lead to field trips and 0.25% to detailed study, and 10 mines have been established as a result. The resourceful Finns also use dogs to sniff out sulphide boulders and so define the dispersal pattern of ore-bearing rocks (Kujansuu 1976).

4.4.5 Geochemistry

The bulk composition of the matrix of till or tillite resembles that of greywackes, which it often is in a mineralogical sense (Pettijohn 1975: 176–8). These materials are normally rich in alumina, iron, alkaline earths and alkaline metals. Tills in limestone and dolostone regions have a matrix that is rich in CaO, MgO and CO_2. Tills deposited in an aqueous environment differ from lodgement and ablation tills because of the effects of sorting by currents and removal of the fines. Consequently, the matrix of such tills is more sandy and thus depleted in Al_2O_3, iron and K_2O, and higher in SiO_2.

4.4.6 Characteristics of quartz and heavy mineral grains

Scanning electron microscopy on quartz or garnet grains enables different transport mechanisms or sedimentary environments to be distinguished on the basis of surface features on sand- and silt-sized particles, many of which are illustrated by Krinsley & Doornkamp (1973). Further details are given in §1.4.5.

4.5 Landforms in terrestrial glacial depositional environments

A wide range of depositional features, notably moraines and ice-contact deposits, are associated with glaciers and ice sheets, many of which are important in elucidating the palaeogeographical and palaeoclimatic patterns of a glaciated region. Their value in this regard is naturally of most significance for Quaternary events, since relicts of glaciation survive at the surface, but the study of glacial and related landforms has proved to be an unreliable method of establishing a glacial chronology. In the earlier geological record such landforms are rarely preserved or at least are unrecognizable. However, there are exceptions, such as in "Ordovician" Sahara and, where landforms are preserved, we have a powerful means of determining palaeo-environmental conditions.

Various classifications of the landforms arising from terrestrial glacial deposition and associated ice-proximal processes have been proposed. Here, we adopt a classification based on the relationship of the landform to its source within or at the margins of the glacier, and to

Table 4.1 Classification of glacial depositional landforms according to position in relation to glacier and ice-flow direction. Glacier surface features before the ice has totally melted are included, but glaciotectonic landforms are excluded.

Position in relation to glacier	Relation to ice flow	Landform	Scale (1 cm – 100 km)
Supraglacial; still actively accumulating	Parallel	Lateral moraine	10 m – 100 km
		Medial moraine	10 m – 100 km
	(Transverse)	Shear/thrust moraine	1 m – 100 m
		Rockfall	1 m – 1 km
	Non-orientated	Dirt cone	10 cm – 1 m
		Erratic	10 cm – 1 m
		Crevasse filling	10 cm – 100 m
Subglacial during deposition	Parallel	Drumlin	10 m – 10 km
		Drumlinoid ridge	100 m – 10 km
		Fluted moraine	10 m – 1 km
		Crag-and-tail ridge	10 m – 10 km
	Transverse	De Geer (washboard) moraine	10 m – 1 km
		Rogen (ribbed) moraine	10 m – 1 km
	Non-orientated	*Ground moraine:*	
		till plain	1 km – 100 km
		gentle hill	10 m – 10 km
		hummocky ground moraine	1 km – 10 km
		cover moraine	1 km – 10 km
Supraglacial during deposition	Parallel	Moraine dump	10 m – 1 km
	Non-orientated	Hummocky (or dead ice/ disintegration) moraine	100 m – 100 km
		Erratic	1 cm – 10 m
Ice marginal during deposition	Transverse	*End moraines:*	
		terminal moraine	10 m – 100 km
		recessional moraine	10 m – 10 km
		Annual (push) moraine	10 m – 1 km
		Push moraine	10 m – 10 km
	Non-orientated	Hummocky moraine	10 m – 10 km
		Rockfall	1 m – 10 km
		Slump	1 m – 10 km
		Debris flow	1 m – 10 km

130

the direction of ice-flow. The classification incorporates shape, sediment assocations and structures, and is based to some extent on a scheme published by Sugden & John (1976) and on a classification generated by INQUA (Goldthwait 1989) (Table 4.1). However, it should be borne in mind that the specific landforms described belong to a complete spectrum of forms. A particular feature may not therefore necessarily fit exactly into any one of these categories. Excluded from this discussion is a consideration of glaciotectonic landforms which are formed from a combination of erosion, folding and faulting, and deposition of bodies of bedrock and sediment. These forms are discussed in §4.6.

"Moraine" has been variously defined but here we follow Sugden & John (1976: 214) in defining it as an accumulation of glacial or glacier-worked sediments having an independent topographical expression. A moraine usually is made up of till, but there are exceptions, such as where ice-push has occurred

A number of excellent texts provide comprehensive summaries of the various glacial depositional landforms (e.g. Embleton & King 1968, Flint 1971, Sugden & John 1976), hence this brief account is intended simply to convey an impression of the wide range of features associated with such environments.

4.5.1 Debris forms on the glacier surface

In a geomorphological sense, these forms are largely ephemeral since, although they are the result of deposition, they have not finally been deposited. Nevertheless, they are striking elements of the glacial terrestrial environment.

Lateral moraines on the surface of a glacier are the result of intermittent rockfall on to marginal ice from ice-eroded cliffs. Since active glaciers tend to be convex in cross section, debris can accumulate in a well defined band between the rock wall and the ice, where it will be subject to grinding and comminution during glacial movement. Thus, lateral morainic debris commonly shows characteristics of both subglacial and supraglacial debris. Many lateral moraines, particularly those associated with cold glaciers, are ice-cored. These typically comprise a debris layer 2 m thick, resting on inactive glacier ice. Lateral moraines may facilitate access to the middle of a glacier that is heavily crevassed at its margins. Although loose, unstable and uncomfortable to walk on, the boulders frequently bridge the crevasses and the debris cover encourages the ice to melt back at lesser angles.

Rockslides are common in steep mountain terrain because of the oversteepened slopes and severe frost action. Earthquakes in some areas, e.g. southern Alaska, New Zealand, Andes, Caucusas, Himalaya, promote instability and rockslides have been known to blanket entire segments of a glacier. Post & La Chapelle (1971) illustrated one such rockfall on Sherman Glacier arising from one of the most powerful earthquakes ever recorded (in 1964). This blanketed the ice to such an extent that it retarded ablation and caused the glacier to re-advance. Another rock avalanche is described in more detail in §2.10.1. Most rockslides form lobes of angular debris extending across the glacier perpendicular to flow. With time they become deformed as they pass down-glacier.

Large isolated angular blocks of rock, sometimes the size of a small hut, may fall into the

Figure 4.13 Medial moraines on the surface of a valley glacier, north of Nordvestfjord, East Greenland. These are derived from lateral moraines where flow units combine.

Figure 4.14 The snout of Gåsbreen, Hornsund, Spitsbergen, showing: (i) medial moraines emerging from beneath the ice having followed englacial transport paths until this point, as a result of early burial in the accumulation area; (ii) terminal-moraine complex with a core of dead glacier ice.

ice surface. Their size is dependent on the bedding and jointing characteristics of the rock. Igneous and high-grade metamorphic rocks with well spaced joints tend to form the biggest blocks, whereas well stratified or fissile rocks rarely produce large boulders. Once deposited, far from their souce, these blocks are called **erratics**.

Medial moraines occur on active glaciers, sometimes extending through to the bed, particularly where the debris originates from a spur between two valley glaciers (Fig. 4.13). Such debris will generally be angular and frost-shattered. Medial moraines may not appear on the surface until the snout is approached (Fig. 4.14), reflecting either early burial of the debris in the accumulation area or the presence of a rock mass below the ice surface in the ablation area. Medial moraines usually form elevated ridges on the glacier because the debris preferentially protects the ice from ablation, but a thin patchy cover of debris might occupy a linear depression because of enhanced ablation when scattered debris is in contact with ice. Most medial moraines are straight, but they may become folded owing to compressive flow if the ice spreads out laterally as a piedmont glacier. Contorted (looped) moraines result from glacier surging (Fig. 2.28). Near the snout of an alpine glacier, medial moraines tend to merge with lateral moraines, forming a complete debris cover a metre or more thick. This may sometimes be stable enough to allow small plants or even trees (in Alaska) to grow on the slow-moving or stagnant ice.

Linear ridges of debris on the glacier surface may also result from ice deformational processes, especially shearing or thrusting of the debris-rich basal layers towards the surface in zones of longitudinal compression. These **shear zones** or **thrust ridges** may extend for several metres, are often arcuate, dipping up-glacier if transverse, or towards the middle of the glacier if parallel to the sides. Such debris has a clear basal imprint including striated stones. The actual mechanism of debris incorporation is disputed, but simple "movement of debris up shear planes" as originally proposed by Goldthwait (1951) has been considered by some to be untenable (Weertman 1961, Hooke 1968). A combination of basal freezing-on of debris, and movement of debris-rich ice along flowlines (Hooke 1968), or along thrusts developed along pre-existing structural weaknesses in the ice (Hambrey & Müller 1978), is more likely for steady-state glaciers, particularly if the relatively thin terminal part of a glacier is frozen to its bed, as in the case of a typical cold glacier (§2.10.3). Thrusting, with displacement rates of $0.1 \, \text{mhr}^{-1}$, was observed near the margin of Variegated Glacier during its 1982–3 surge, so it is likely that debris-bearing basal ice can be conveyed along thrusts to the surface if strain rates are sufficiently high (Sharp 1985a).

Another form of debris mound is the **dirt-cone,** which can occur anywhere on the glacier surface, most commonly on otherwise bare ice in the ablation area or on snow. The dirt-cones result from debris accumulating in pools in supraglacial streams; diversion of drainage followed by ablation combined with the protective influence of the debris cover, allows a mound to grow. Dirt-cones can attain several metres in height, and once formed may be relatively persistent features on the ice surface. Even so, the veneer of debris is superficial, usually no more than a few centimetres thick. Dirt-cones on snow are less permanent; they result from slurries of debris flowing on to a glacier during the spring melt.

Crevasse-fillings are features with a consistent orientation and they represent zones where debris was washed into an otherwise clean crevasse by surface meltstreams. They are characterized by their parallelism with crevasse traces (§2.8.3). A special type of crevasse-filling is associated with surge-type glaciers as on Eyabakkajökull, Iceland (Sharp 1985a,b) These really represent the intrusion of dykes of finer material from subglacial diamicton into crevasses as the glacier sinks to its bed after a surge.

4.5.2 Landforms formed subglacially, parallel to ice-flow

Landforms that form at the glacier bed parallel to the direction of ice-flow are streamlined, and probably reflect a complex interplay between deposition and erosion of the unconsolidated sediment. These "bedforms" provide key evidence for understanding the processes that operate beneath glaciers and ice sheets, and are useful in palaeo-environmental reconstruction. However, the processes responsible for the creation of subglacial bedforms are still not well understood, and they have been the subject of major controversy. The debate has centred around: (a) those who advocate subglacial bed formation resulting from the deformation of soft sediments by moving ice; and (b) those who suggest that subglacial flooding on a catastrophic scale over specific areas of the substrate is responsible. Although it could be argued that these bedforms are erosional features, they are included in this chapter since they comprise unconsolidated sediments that may eventually be preserved in the stratigraphic record.

Drumlins are one of the most distinctive features of glacial depositional environments, and are particularly common where broad valley glaciers or ice sheets were flowing relatively fast. Arguably, they are the most intensively studied glacigenic landform, partly because of their value in reconstructing ice-flow directions, but more especially because of the light they ought to throw on processes operating beneath large ice masses. Major contributions in the field of drumlin research in recent years are contained in the volumes edited by Menzies & Rose (1987, 1989), while Embleton & King (1975) have given a thorough review of earlier work with many examples. The term drumlin is derived from *druim*, a Gaelic term for a mound or rounded hill.

Drumlins come in a great variety of shapes and sizes, for example, ellipsoidal, egg-shaped, and irregular multiple ridges. The degree of elongation is variable (normally in the range 2.5:1 to 4:1, although exceptionally as much as 60:1) and often they have a long low tail. They occur singly, or in fields of hundreds, whence they give rise to the expression "basket of eggs" topography, on account of their resemblence to birds' eggs, which have a prominent asymmetry with the blunt end facing upstream. The larger drumlins reach 50m or more in height and 20km in length, but others might be only 2m high and 10m long. Drumlins are part of a continuum of glacier bedforms that extends into flutes (Rose 1989).

Drumlins are widely, but sporadically developed in the depositional zones of the last great ice sheets of the Northern Hemisphere. The largest drumlin field, containing some 10000 individuals, is probably that in central-western New York State. It occupies an area of 225km by 56km between Lake Ontario and the Finger Lakes, and was deposited by the Wisconsinan Laurentide ice sheet. In the British Isles, extensive drumlin fields occur in southern Scotland, northwest England, parts of Wales, and in northern and western Ireland (Fig. 4.15); these were formed by the main Devensian ice sheet. Elsewhere, in Europe the extensive drumlin

Figure 4.15 Drumlin field, SE Cumbria, England, *viewed* towards the Howgill Fells (left background) and the Pennines beyond and to the right. Ice-flow was towards the right. Note the steeper up-glacier ends of many drumlins (in shadow) (photograph: BPE–86 with permission of Cambridge University Collection of Air Photographs, copyright reserved).

fields deposited by Weichselian ice sheets occur in Sweden and Finland, while smaller fields, related to the encroachment of Würmian glaciers on to the lowlands, are to be found in Switzerland and Germany.

The composition of drumlins is highly variable. Usually the dominant facies is massive diamicton, interpreted as till. Some observations have shown that the preferred orientation of clast long axes is parallel to the drumlin long axis, and hence to ice-flow, but in some cases the preferred orientation varies throughout the drumlin, although tending to point towards the long axis. Some drumlins have a rock core and others comprise coarse, stratified glaciofluvial deposits. These sediments are commonly highly deformed, with folds overturned in the direction of ice-flow; in such cases subglacial bed deformation is likely to be the dominant process.

A number of hypotheses have been proposed to account for drumlins, according to a variety of subglacial conditions (Menzies 1989).

(a) Moulding of previously deposited material within a subglacial environment in which a limited amount of subglacial meltwater activity occurs.

(b) Formation resulting from textural differences in subglacial debris due to dilantancy (ability to contract in volume), pore-water dissipation, localized freezing, or localized basal ice-flow patterns.

(c) Formation due to the effects of active basal meltwater carving cavities beneath an ice mass and subsequently infilling the space with a variety of stratified sediments, or by meltwater erosion of *in situ* sediments. Some of the major drumlin fields (e.g. Livingstone Lake, Saskatchewan) have been explained as the product of catastrophic meltwater floods beneath the large Pleistocene ice sheets, since the pattern and geometry of these features bear a close resemblance to bedforms such as beds of ripples or barchan dunes, which are the product of turbulent water-flow in other situations (Shaw et al. 1989).

Whichever of these mechanisms is correct, there is increasing evidence to suggest that deformation of unconsolidated, slurry-like material beneath a glacier is an important process (§2.7.3), Beneath fast-flowing glaciers, localized perturbations within the slurry may eventually lead to the development of at least some groups of drumlin-like landforms.

Drumlinoid ridges or **drumlinized ground moraine** are elongated, cigar-shaped ridges, and spindle forms (Fig. 4.16). Like drumlins, they are the product of ice streamlining, but under basal ice conditions that were unsuitable for discrete drumlins to form.

Fluted moraines or **flutes** represent an end member of streamlined forms that form on fresh lodgement till surfaces. However, flutes may occasionally appear on surfaces of other material, for example, glaciofluvial sand and gravel. Generally they appear as large furrows about a couple of metres in wavelength, reminiscent of ploughed ground (Fig. 4.17), but megaflutes 100 m wide, 25 m high and 20 km long have been reported in Montana, USA. Superimposed flutes on megaflutes have been described from the proglacial area of Austre Okstindbreen, northern Norway by Rose (1989). Many flutes extend down stream of large boulders embedded in lodgement till. There is usually a moderately strong preferred orientation of clasts parallel to the flute, although there may be some divergence from the linear trend on the flanks of the flute. Embedded boulders may attain an abrupt stoss side and gentle downstream form (Boulton 1978). Many mechanisms have been suggested, but most authors

Figure 4.16 Drumlinoid ridges and fluting, south of Thelon River, District of Keewatin, Northwest Territories, Canada. Ice-flow was towards the bottom of the picture (photograph NASPL T301L–223 with permission of the Department of Energy, Mines and Resources, Canada).

accept that at least some flutes are formed by the squeezing of a saturated till into the hollow formed by ice as it moves over a large boulder embedded in the till.

Boulder beds with striated upper surfaces (**boulder pavements**) are massive, matrix-supported diamicts with moderate to poor sorting. Gravel clasts are mostly rounded and subrounded, and have long axes preferentially aligned parallel to the striae on the top surface of the pavement. Lodgement of basal debris, incorporated into the glacier from previously winnowed till, boulder beaches or fluvial deposits, is thought to be responsible for the boulder beds. Selective lodgement of boulders occurs down-glacier when basal thermal conditions change from cold/freezing to warm/melting. Excellent examples have been described from the Late Palaeozoic Dwyka Formation of South Africa (Visser et al. 1987).

4.5.3 Landforms formed subglacially, transverse to flow

The origin of transverse moraines formed subglacially has received much less attention than longitudinal streamlined forms, even though they are often closely associated. Transverse

137

Figure 4.17 Fluted moraine in the proglacial areas of the glacier Austre Okstindbreen, Okstindan, northern Norway. The glacier snout lies to the right of the proglacial lake.

moraines have a variety of names reflecting those of their discoverers, type locality or mode of formation.

Rogen or **ribbed moraines** are large-scale transversely orientated, somewhat irregular, ridges, typically 10–20 m high, 50–100 m wide and 1–2 km long. They were named after a lake in Sweden, and have been recently reviewed by the originator of the term, Lundqvist (1989). The ridges are often slightly arcuate and concave up-glacier. Frequently, irregular cross-ribs link up three or four of the transverse ridges, enclosing small lakes and boggy hollows (Fig. 4.18). One of the most characteristic features of the ridges is their gradual transition into drumlins.

The moraines are composed of clast-rich diamicton and sediments laid down by water. Commonly, a collection of large boulders sits on top. Measurements of the long axes of clasts indicate a preferred transverse orientation. The association of Rogen moraines with stream-lined sediments is illustrated, for example, by their crests being fluted or drumlinized. Rogen moraines may also pass imperceptably into drumlins in a down-glacier direction.

The following hypotheses for the origin of Rogen moraines have been discussed by Lundqvist (1989).

(a) Deposited as marginal moraines, forming a complex of end moraines.

(b) Deposited as subglacial moraines, as emphasized by modern literature, formed under thick ice away from the ice front, especially in a transitional zone between warm- and cold-based ice where the ice is under compression.

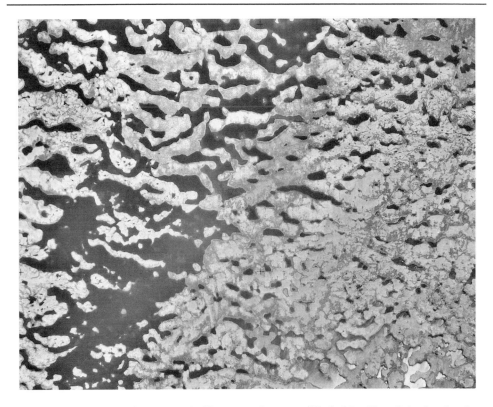

Figure 4.18 Vertical aerial photograph of Rogen moraine, west of Kaniapiskau River, Labrador, Canada; the ridges are 15–20 m high (photograph NAPL A11441–121, with permission of the Department of Energy, Mines and Resources, Canada).

(c) Formation by active ice, as a result of tectonic processes within the ice, i.e. subglacial folding of debris-rich layers, or stacking of thrust-slices of debris-rich ice against obstacles to glacier flow, followed by melt-out.

(d) By filling of open crevasses by supraglacial debris.

(e) By filling of basal crevasses by subglacial debris.

Lundqvist's most favoured hypothesis for the formation of Rogen moraines is a combination of some of these processes, and he notes a genetic affinity with drumlins. In areas of extending flow near the ice margin, drumlins form at obstacles to the flow. Where the flow becomes compressive, drumlin formation is incomplete, and only crescent-shaped ridges form. These forms are preserved only in the central part of a glaciated area, where the ice stagnated. Stagnant ice deposits are often superimposed on Rogen moraines. The importance of shearing/thrusting mechanisms up-glacier of the snout in the creation of Rogen moraines has been stressed by Sugden & John (1976) and Bouchard (1989).

Transverse moraines that are the product of **surging** have been described from Eyabakkajökull in Iceland by Sharp (1985b). Here, broad, sedimentologically complex ridges up to 25 m high are present. The processes held to be responsible for their formation include: thrusting, gravity-flow of glacigenic sediment, debris-dyke formation, glaciofluvial activity and wastage of buried stagnant ice. Moraine complexes of surge-type glaciers often appear as chaotic, hummocky moraines with many kettle holes between, but inspection of aerial photographs of such areas often reveals a clear linear trend to the hummocks (Fig. 4.19). In North America a hummocky supraglacial complex extends over half a million square kilometres in Alberta, Saskatchawan and Manitoba. Here the relief of these features attains 100 m (Paul 1983). Towards the southern limit of the Wisconsinan (last) ice sheet, it is conceivable that hummocky moraines were formed following a surge of part of the ice sheet over the Mid-West United States. Rapid down-wastage followed the advance (however caused) and blocks

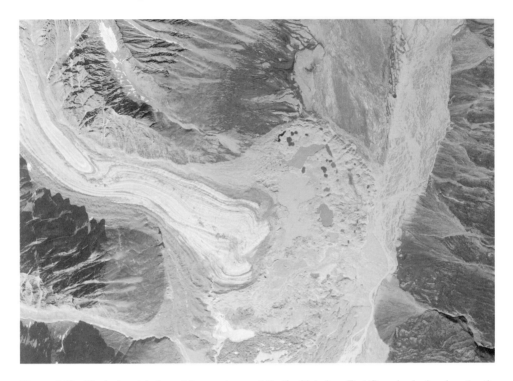

Figure 4.19 Vertical aerial view of the snout area of Roslin Gletscher, East Greenland, showing chaotic ice stagnation features formed following a surge. Hummocky moraine with intervening lakes, and thrust moraines are the dominant features, and show a weak transverse trend (photo 278C no. 334, with permission of Kort-og Matrikelstyrelsen, Boligministeriet, Copenhagen).

of ice may have become detached from the parent mass. Certain areas show superimposed transverse ridges which could represent crevasse-fillings, creating an assemblage of forms that is characteristic of surge-type glaciers in Iceland as described by Sharp (1985b). However, this concept of surge-type deposition for the Mid-West moraines is highly controversial.

Other types of transverse moraines include De Geer and sublacustrine moraines; they are discussed in Chapter 6.

4.5.4 Non-orientated landforms, formed subglacially

Landforms of this nature are commonly referred to, somewhat loosely, as ground moraine – areas of non-lineated, smooth to hummocky drift cover, especially of basal melt-out and lodgement till, with minor subglacial outwash lenses and a thin cover of supraglacial melt-out or flow tills. The various types of ground moraine, from thickest to thinnest are:

- **till plains** – nearly flat or slightly rolling and gently inclined plains, mostly consisting of a thick till cover, often in multiple layers of varying composition, completely masking bedrock irregularities;
- **gentle hills** – mounds of till resting on detached blocks of bedrock;
- **hummocky ground moraine** (or **dead ice/distingegration moraine**) – a chaotic area of hills of basal till, flow till with minor reworked sediments associated with small streams and ponds (supraglacial debris may drape the basal till);
- **cover moraine** – a patchy, thin layer of till revealing the bedrock topography, either entirely (veneer) or only partly (blanket).

Till plains are normally associated with lowland glaciation. Large flat-lying areas underlain by till are widespread in Europe and North America. Much of the Cheshire Plain, the Vale of York, and East Anglia in Britain has flat sheets dominated by till, often 100 m thick. The plains of North Germany and lowlands north of the Alps have a mantle of till. In North America, till plains are widespread in the states of the Mid-West. Hummocky moraines are common in both lowland and mountain regions and are a form of ablation moraine. Hummocky

Figure 4.20 Late glacial hummocky moraines in Glen Derry, Cairngorms, Scotland.

moraines are the result of down-wastage of a glacier rather than recession, especially when the ice has ceased to be active. They are often genetically linked to supraglacial hummocky moraines, where basal debris is elevated to an englacial and supraglacial position, and mixed with supraglacial debris. It is normal for the hilly topography to be controlled by supraglacial melt-out processes, rather than basal melt-out, so these forms are discussed more thoroughly in §4.5.6. One of the most instructive papers on the genesis of these features was by Boulton (1967), who described processes at the margins of a glacier in Spitsbergen.

Hummocky moraine is a characteristic feature of the southern limits of the last great ice sheets of North America and northern Europe, where they attain a relief of tens and, occasionally, a few hundred metres. Large areas of hummocky moraine suggest that the ice sheets wasted rapidly. It is unlikely that supraglacial debris was much involved; the only potential source is from the bed by means of shear processes anyway, and such material is indicative of active ice, and generally limited to the snout position. In the British Isles, one of the largest supraglacial hummocky moraine complexes was formed when a major ice stream flowing down the Irish Sea impinged against the west flanks of the Pennines in Cheshire and Staffordshire (Paul 1983).

Hummocky moraines in highland areas are a feature of down-wastage of valley glaciers. They can be observed to form in almost any terrestrial glacial environment today. Particularly fresh-looking forms, with a relief of several tens of metres in valleys of the English Lake District and the Scottish Highlands, were deposited during the rapid wastage phase of the last valley glaciers and ice fields around 10000 years BP (Fig. 4.20).

Some erratics belong to this class of landform. Where large blocks have been detached from the bed of a glacier they may stand proud of the general level of the till plain.

4.5.5 Supraglacial landforms, parallel to ice-flow

Features of this nature include lateral moraines and moraine dumps, the latter derived from medial moraines; these have already been mentioned in the context of features present on the surface of moving ice. However, the long-term preservation potential of lateral moraines is relatively small due to valley-side collapse after ice-recession. That of medial moraines is even less because of ice-front processes and glaciofluvial action.

Lateral moraines are among the most impressive features of contemporary glacial mountain environments (Fig. 4.21), especially above and down stream of those glaciers in the Alps, Scandinavia, the Western Cordillera of North America, and elsewhere which advanced strongly during the Little Ice Age of around AD 1700–1900.

Lateral moraines form from the steady supply of frost-shattered debris that falls down on to the glacier margin from the cliffs above, and from debris that is rubbed between the glacier and the valley side. They are, therefore, composite supraglacial and subglacial features. The debris reflects both the angular nature of rockfall debris and the subrounded/subangular nature of basally (or marginally) transported stones. The latter type often show typical glacial abrasional markings, such as striations and faceted surfaces. The effect of comminution of debris is such that lateral moraines generally have a mud-grade matrix. This is manifested in

Figure 4.21 A fine pair of lateral moraines along the flanks of Vadrec da Tschierva, Graubunden, Switzerland. Note the extent to which material has been accreted away from the valley sides, and the characteristic small valleys between the moraine crests and adjacent hillsides.

the steep inner faces of lateral moraines when ice recedes rapidly, the clay binding the slope to some extent together, although the effect of rainwater and irregular dislodging of embedded boulders tends to produce parallel runnels (Fig. 4.21). It is deceptively difficult and often hazardous to climb the inner faces of fresh lateral moraines. In contrast the outer face of a lateral moraine, especially a Little Ice Age one, is stable and often well vegetated. This slope, facing the adjacent hillside, may have its own stream and small lakes at its foot.

Lateral moraines in Alpine terrain often grow towards each other by accretion of basal or marginal, rather than rockfall debris. In places, such moraines may become somewhat detached from the valley sides (Fig. 4.21). The glacier may thus be more constrained by its moraines than by the steep valley sides. Advancing glaciers may break through the old moraines below the crest or dislodge boulders over the crest.

Lateral moraines that cling to the mountainside rapidly undergo collapse. After a few hundred years, a low-angle bench or a slight break in slope may be all that is left of a lateral moraine. However, even in Britain, which lost its glaciers 10000 years ago, a discerning eye may be able on hillsides to pick out subtle notches that represent lateral moraine positions.

Another type of lateral moraine is ice-cored, and is a feature that is best developed in the polar regions where melting of a large mass of ice takes hundreds of years, although some good examples occur in Alpine terrain (Fig. 4.22). Ice-cored moraines may also comprise

Figure 4.22 Lateral moraines alongside Glacier de Corbassière, Valais, Switzerland. The outermost one (left) is probably the Little Ice Age moraine; the other three are ice-cored.

basal and supraglacial debris. Dynamically, they are separate from the main glacier; in fact, the ice is probably dead. The veneer of debris is typically 1–2 m thick, and dark wet scars indicate where slumping off the sides has occurred. Alongside cold glaciers, marginal streams may flank both sides of the moraine, one associated with glacial meltwater, the other with snow melt and rainwater.

Moraine dumps rarely survive the recession of a glacier. Near the snout, the debris is often spread over a broad zone, and several medial moraines merge together to form a blanket, rather than distinct ridges. Even if a prominent medial ridge is left when the ice recedes, glaciofluvial action will tend to destroy it as the river braids constantly migrate across the valley floor.

4.5.6 Non-orientated landforms, formed supraglacially

As mentioned in §4.5.4, **hummocky moraines** are often of composite origin, involving the displacement of basal debris towards the glacier surface. However, in contrast to lowland ice sheets and cold glaciers, which yield hummocky moraines of largely basal origin, mountain glaciers generate topographically similar features principally as a result of down-wastage

144

of a thin supraglacial debris cover, derived mainly from rockfall.

Hummocky moraine begins as unstable debris lying on a substantially debris-covered glacier surface, usually near the snout where the ice is relatively inactive. Uneven ice wastage gives rise to a chaotic hummocky appearance, and recycling by flowage of saturated debris and by local supraglacial and englacial meltwater streams is a common process. Sometimes, meltwater has carved out of the ice large tunnels and caverns, an assemblage of forms referred to as glacier karst by analogy with limestone regions. The relief in areas of hummocky moraine may reach several tens of metres. The process of hummocky moraine formation may be observed directly on many glaciers, for example, the Unteraargletscher in Switzerland and the Tasman Glacier in New Zealand. Ultimately, the dead glacier ice is reduced to areas of ice-cored moraine and the final decay of ice leaves behind a series of hummocky moraines.

Mention should be made of supraglacially derived erratics. These angular blocks may have fallen on to clean ice and may not be linked with any surface moraine, so are isolated when the ice recedes. Occasionally, such erratics stand on pedestals if the underlying surface has since been eroded away, especially in areas of carbonate bedrock, such as the Pennines of northern England, where chemical weathering has taken place.

4.5.7 Ice-marginal landforms, transverse to flow

These forms are essentially represented by a range of **end moraines** that document the stages when a glacier remained stationary but active, or more especially when it advanced. They form perpendicular to ice motion and consist of belt of ground higher than the general level of the valley floor, often sweeping in a convex down-valley arc to join lateral moraines. End moraines are compositionally varied and complex, comprising any of the sediments that form in the proglacial area of a glacier.

A **terminal moraine** is a broad arcuate belt of debris that formed around an ice lobe during the period in which the ice was at its maximum extent. It is frequently hummocky, pitted, comprises irregular short ridge crests, and contains high mounds of distinctive lithology derived from point sources. A long halt during a general phase of recession leads to the formation of a **recessional moraine**.

Both these sorts of moraine have a steep outerface when fresh (20°–30°), whereas the inner slope is generally irregular and hummocky, and of relatively low angle (10°–20°); they tend to have a convex down-glacier form in valleys, and even in lowland areas they are frequently lobate. In regions such as the Alps, well preserved terminal moraines are relatively rare, owing to destruction by meltwater. Lower land bordering the Alps and similar regions has well formed moraines, deposited by glaciers of the last ice age, notably those which enclose the outlets of the large lakes in northern Italy. In North America, a fine set of terminal moraines lies to the east of the Teton Range in Wyoming. These were created by a piedmont glacier that spread out from the foot of the range and scooped out the lake basin which is now enclosed by a prominent ridge. One of the tallest terminal moraines (430 m) was created by the highly dynamic Franz Josef Glacier when it debouched on to the coastal plain to the west of the Mount Cook range in New Zealand. Terminal moraines associated with the great ice sheets do not have such pronounced relief, but are impressive because of their lateral extent; northern Germany, Poland and Denmark have fine examples. The southern margin of the Laurentide ice sheet in North America is marked by a zone up to 600km wide, comprising a terminal–recessional moraine complex, extending across the continent to the Rockies in the latitude of, and south of, the Great Lakes. North of this zone is a belt of ice-disintegration features, the hummocky moraines, to which reference has already been made.

Dumping of supraglacial debris is a common mechanism in the creation of terminal and recessional moraines, since during an advance the snout is normally very steep and debris accumulates as a heap at the foot of the ice slope. Release of the basal debris load in one place for a long period will result in an irregular terminal moraine composed principally of subangular and subrounded stones, of which a proportion will be striated. The process of thrusting, already described, may create distinct ridges. This latter mechanism is important at the margins of cold glaciers in polar regions today, and the process may have been significant in the creation of the end moraines of the last glaciation.

Other processes involved in terminal moraine formation include melt-out, flowage and squeezing out of saturated till. Ice-cored end moraines may also develop, especially in the polar regions (Fig. 4.14). Clast fabric studies have tended to indicate a preferred orientation parallel to the moraine crest. Terminal and recessional moraines are often destroyed by meltwater during subsequent recession, and those that survive may have relatively subdued forms. Alternatively, they may simply be represented by boulder belts, scattered out-size boulders (up to the size of a hut), the small debris having been washed away.

Rather more ephemeral features are **annual push moraines**, which are the result of small winter re-advances pushing up a series of small, closely spaced ridges a metre or so high, during a general period of recession. They are a common feature at the snouts of the relatively less dynamic temperate glaciers of Norway, and of smaller glaciers in alpine areas (Fig. 4.5).

The rather special forms of end moraine that result from glaciotectonic processes are described in §4,6. However, it is conceivable that many more, supposedly "conventional" end moraines may turn out on further investigation to be the product of glaciotectonism.

4.5.8 Non-orientated ice-marginal and proglacial landforms

Non-orientated ice-marginal and proglacial landforms include hummocky moraines and the product of mass movement of glacigenic sediment. The genesis of hummocky moraines is normally associated with supraglacial or basal melt-out processes (as described in §4.5.4), but continues if irregular heaps of debris fall from the face of an ice mass into the ice-marginal zone, or if dead ice becomes detached from the parent ice body.

Mass-movement processes which result in discrete landforms are rockfalls, debris-flows and slumps. Rockfalls are induced by (a) the oversteepening of valley sides by glacial erosion and (b) the development of joints parallel to the valley side as a result of pressure release following recession of the ice (Harland 1957), a type of exfoliation. Such rockfalls can sometimes fill a valley bottom, and many instances occur in alpine valleys. Debris-flows and slumps most commonly involve tills. The high silt and clay content allows till-covered hillsides to fail easily, especially when wet. Debris-flows commonly occur as distinct lobes, often several metres thick. Slumps are marked by scars of fresh debris and a rucked surface and bulge below. These processes are a day-to-day occurrence on land recently vacated by ice, and may continue to affect hillsides thousands of years after the ice has receded.

4.6 Glaciotectonic landforms

Glaciotectonic structures are commonly manifested on the land surface as distinct landforms. Many late Quaternary and younger features display their original morphology, but others may have been buried by later depositional processes. Glaciotectonic landforms comprise a variety of hills, ridges and plains, all of which are constructed wholly or in part of bedrock or drift, or both. The definitive work on glaciotectonic landforms (Aber et al. 1989) uses a five-fold classification. This is described below following the order of decreasing topographic prominence.

4.6.1 A hill–hole pair

A **hill–hole pair** represents a combination of an ice-scooped basin and a hill of ice-thrust, often slightly crumpled, material of similar size. The principal morphological features include:
(a) an arcuate or crescentic outline of a hill, convex in the down-glacier direction;
(b) multiple, subparallel, narrow ridges following the overall arcuate trend of the hill;
(c) an asymmetrical cross profile, with steeper slopes on the down-glacier side;
(d) topographic depression on the up-glacier side of the hill, covering an area approximately equal to that of the hill.

The sizes of the hill–hole pairs vary from 1 to 100 km^2, with relief ranging from 30 to 200 m. Good examples of hill–hole pairs are situated at Wolf Lake in Alberta, Herschel Island in the Yukon and at the contemporary ice margin of Eyabakkajökull in Iceland.

4.6.2 Large composite-ridges

Large composite-ridges are the most typical and distinctive of all glaciotectonic landforms and are composed of large slices of up-thrust and commonly contorted sedimentary bedrock that is generally interlayered and overlain by large amounts of glacial drift. Large composite-ridges are up to 200 m high, 5 km wide and up to 50 km long, and have an arcuate form. Individual ridges within these complexes are typically several hundred metres high and up to 100 m wide. Elongated lakes often form in the valleys between the ridges. The inside of the ridge complex generally demarcates the maximum position of the ice. Large composite-ridges usually involve considerable disruption of pre-Quaternary bedrock, perhaps to depths of 200 m

147

below the surface. The folds and thrust-blocks are stacked up in piggy-back fashion. Large composite-ridges are geometrically similar to mountain belts that were the product of thin-skinned tectonic movements.

The best known site is Møns Klint in Denmark, where large thrust-blocks of Cretaceous chalk and drift are exposed in coastal cliffs up to 143 m high, with ridges extending inland. This complex was formed during the late Weichselian glaciation, as a result of the expansion of the Fennoscandian ice sheet. Similar ridge complexes have been described from the Dirt Hills and Cactus Hills, Saskatchewan, and the Prophets Mountains in North Dakota.

4.6.3 Small composite-ridges

Small composite-ridges are smaller scale versions of the ridges described above, and they generally have a relief of less than 100 m. Many are composed only of unconsolidated Quaternary strata and are thus more susceptible to erosion. The terms push moraine and *Stauchmoränen* (from the German) are commonly applied to such ridges. They are the product of thrusting of unconsolidated proglacial deposits, including glaciofluvial and marine sediments, especially in areas of permafrost. Good examples are found in the Canadian Arctic and Svalbard. On Axel Heiberg Island, the snout of advancing Thomson Glacier is pushing forwards a small composite-ridge 45 m high, 2.1 km long and 0.7 km wide, made up almost entirely of glaciofluvial sediments (Kälin 1972) (Fig. 4.23).

Figure 4.23 Advancing front of Thomson Glacier, Axel Heiberg Island, Canadian Arctic. To the right a push-moraine complex (or "small composite-ridge") of thrust fluvial gravels is forming. This complex was described in detail by Kälin (1971).

The forward part of a push-moraine complex usually comprises stacked sheets of glaciofluvial sand and gravel, forming ridges with steep outer faces and gently inclined inner faces, with no trace of till, so the sediment must have been thrust up in a frozen state in front of the ice margin. Inner parts of the complex comprise sheets of sand and gravel, but also till, glaciolacustrine or glaciomarine sediments. Recumbent folding is evident, especially in the less competent muddy materials.

Composite-ridges seem to be preferentially associated with surge-type glaciers. Many small composite-ridges have been mapped as conventional end moraines, but for these a glaciotectonic origin cannot be ruled out. A continuous spectrum occurs between the two types of composite-ridge.

4.6.4 Cupola-hills

Isolated hills or a jumbled group of hills with no obvious source-depression, but having the general characteristics of ice-thrust masses, occur on both large and small scales. The most common form is the **cupola-hill**, which has an internal structure similar to that of composite-ridges, but which shows signs of having been subsequently overridden by the ice. Cupola-hills consist of deformed glacial and interglacial deposits, plus detached blocks (**floes**) of older strata or bedrock, overlain by basal till which truncates the older strata. Cupola-hills have a dome-like form with long, even slopes, varying from near-circular to elongated ovals. They range from 1 to 15 km in length, and from 20 to at least 100 m in height. Cupola-hills are common in regions having a soft substratum that was affected by ice coming from different directions. Good examples of cupola-hills, described by Aber et al. (1989), occur on the islands of Møn and Langland in Denmark and at Martha's Vineyard in Massachusetts.

4.6.5 Mega-blocks and rafts

Large pieces of glacially transported bedrock buried in drift are called **mega-blocks** or **rafts**. Although they are typically less than 30 m thick, their lateral dimensions often exceed 1 km². Transport distances may be great; for example, a group of rafts in east-central Poland originates in Lithuania, over 300 km to the northeast. Rafts are generally removed from the bedrock along bedding planes, which act as planes of detachment when the ice overrides them, especially if the ground is frozen.

Perhaps the biggest raft, covering an area of about 100 km² is found at Esterhazy, Saskatchewan, and many other rafts occur throughout the plains of southern Alberta. Rafts of Cretaceous chalk, measuring up to about 100 m in length are well exposed in coastal cliffs in Norfolk, England. These show signs of internal tectonic deformation, including thrusting and gentle folding. A dolomite example from the Neoproterozoic Port Askaig Tillite in the Garvellachs, Scotland, and nicknamed "The Bubble", is tightly folded.

149

4.6.6 Diapirs, intrusions and wedges

Various kinds of soft-sediment deformation features are widely known from glacigenic sequences, and range in lateral dimensions from a few centimetres to over 100 m. Intrusions include all those structures resulting from the injection or squeezing of one type of material into another. The intruded material is usually clay- or silt-rich sediment, whereas the host material may be of almost any composition if soft. Intrusions develop in a subglacial, water-saturated condition with intergranular movement as the main means of deformation. Structures may be grouped as those that originate from below (diapirs, dykes, etc.) and those that are forced down from above (wedges and veins). Wedge structures may be misinterpreted as periglacial features, such as ice wedges or thermal crack-fillings. Good examples of intrusions have been described by Aber et al. (1989) from Kansas, western Norway and Denmark.

4.7 Preservation potential of terrestrial glacigenic sediments

Compared with many non-glacial sedimentary sequences, terrestrial glacigenic deposits tend to be relatively thin. In highland areas their potential for preservation is low because of the efficacy of reworking processes, such as those of streams or mass movement, following ice-recession. In less confined lowland areas, sheets of till may have a chance to build up to a thickness of several tens of metres and remain largely undisturbed by non-glacial processes. On the other hand, subsequent ice-advances tend to remove the evidence of the previous depositional phases, and the earlier record of glacier fluctuations will be largely destroyed.

On a geological timescale of tens and hundreds of millions of years, even a somewhat sparse terrestrial record may be preserved from lowland areas, since the decay of ice sheets leads to a rise in sea level, thus burying the glacial sequences with marine sediments. Thus, although unimportant in volumetric terms, terrestrial glacigenic sediments are documented from all the main phases of the Earth's glacial history. Probably the best examples are from the Late Palaeozoic glaciation of Gondwanaland. However, to assess the ancient glacial climatic record, we need to turn to glaciomarine sequences, since it is these that provide the thicker and more continuous sedimentary sequences, and which are less likely to have been eroded.

In contrast to the sediments, glacial depositional landforms do not commonly survive intact to be incorporated in the sedimentary record. Indeed, many moraines and ice-contact features disappear when the glacier itself has gone because they depend on the ice for support and are subject to reworking by mass movement, fluvial action, periglacial processes, etc. Some streamlined forms, in theory, have a better chance of survival in the sedimentary record than some less dynamically produced forms because they may quickly become buried as a result of sedimentation associated with ice wastage. However, to recognize such forms in sedimentary sections can be difficult, and there are only a few instances of depositional landforms being identified in ancient glacigenic sequences.

5 Glaciofluvial processes and landforms

5.1 Introduction

In many areas, more sediment is transferred out of the glacier system by meltwater than directly by the ice. Huge quantities of sediment may be involved, but the rate at which it is moved is strongly dependent upon temperature, which in turn is a function of the season and the time of day. Often, one may obtain a good impression of the power of a meltstream from the noise of boulders rolling along the bed of the stream. The milky appearance of the water arising from suspended mud (**rock flour**) is another indication of the efficacy of such streams in removing sediment from a glacier (Fig. 5.1). Meltwater streams are also powerful agents of erosion, especially beneath the ice where the water may be under high pressure.

The rôle of water within, alongside, and under, the ice, and the character of runoff, has been discussed in Chapter 2. Here, we briefly examine erosional and sedimentary processes and the rôles they play in the development of landforms. Emphasis is placed on processes and landforms in close proximity to the ice margin, rather than in areas distant from the source glacier. For more extensive accounts of glaciofluvial processes and landforms, several works may be recommended: Price (1973), Embleton & King (1975), Jopling & McDonald (1975), Sugden & John (1976), Miall (1977, 1983b) and Drewry (1986).

Many examples of glaciofluvial sequences occur in the geological record. Of particular note are the Neoproterozoic and late Ordovician deposits of the Taoudeni Basin of West Africa (Deynoux & Trompette 1981a,b), and the Late Palaeozoic of the northern Karoo Basin in South Africa (von Brunn & Stratton 1981).

5.2 Glaciofluvial processes

5.2.1 Meltwater erosion

Glacial meltwater at the base of a glacier, together with its sediment load, is often a major instrument of erosion. The importance of meltwater erosion and transport increases progressively towards the snout, thus assuming a dominant rôle below the equilibrium line. Water at

Figure 5.1 The Copper River, Alaska, one of the world's largest glacial meltwater streams. Laden with fine-grained glacial sediment, the stream has a milky appearance as it flows past and undercuts the terminus of Childs Glacier.

the base is primarily derived as a result of supraglacial melting, reaching the bed via crevasses and moulins (§2.5.2). Some water is derived as a result of basal melting and some as runoff from the valley sides, but generally the latter components are much smaller.

Water at the base of the glacier flows in four main modes:
(a) in tunnels cut into the ice, manifested at the snout by a semi-circular or semi-elliptical basal stream outlet or portal;
(b) by sheet-flow at the ice/bedrock interface;
(c) by draining into the underlying sediment, allowing it to deform more easily;
(d) in tunnels incised into the bed, including both bedrock and unconsolidated sediment.

Water is often under high pressure at the base, so facilitating erosion of the ice as well as the bedrock or sediment beneath. The hydroelectric power companies have long recognized the problem of water under high pressure in tunnels, and the process of **cavitation**, i.e. the growth and collapse of air bubbles, resulting in the generation of shock waves. The same effect is evident beneath glaciers, but is compounded by the abrasive quality of suspended sediment.

Meltwater stream patterns are, to a great extent, related to the temperature of the ice. In temperate glaciers, most water finds its way to the base and appears at the snout near the lowest point. Lateral meltstreams are absent. In cold glaciers the meltwater path is dependent on whether the glacier is frozen to the bed, but most water flows near the surface, with discharge taking place via lateral meltstreams, subglacial meltstream systems normally being poorly developed.

Ice-dammed lakes, on their release, also play a significant rôle in meltwater erosion. Typically, they build up during the ablation season until a critical level is reached. This may occur when the water has a sufficient head to force its way beneath the glacier, and to lift up the ice bodily, thus releasing a catastrophic flood of water, which flushes out the system. Alternatively, the lake may overflow and cut a gorge into the glacier surface or spill over a neighbouring col. Subglacial lakes have also been recorded. In sedimentological terms, the consequences of catastrophic drainage from ice-dammed and subglacial lakes are frequently dramatic.

5.2.2 Transport of sediment and solutes

Sediment carried by meltwater streams may be categorized as suspended load and bedload. Suspended load has been monitored near the snouts of many glaciers, and sometimes a clear relationship of major peaks with discharge has been demonstrated. For example, the glacier Erdalsbre in southwestern Norway indicated peaks of suspended sediment concentration coinciding with high discharge events (Østrem 1975). Peak values of sediment concentration exceeded $2000\,\mathrm{mg\,l^{-1}}$ at 10–$15\,\mathrm{m^3\,sec^{-1}}$, whereas typical low values for the same summer season were 100–$200\,\mathrm{mg\,l^{-1}}$ and $5\,\mathrm{m^3 sec^{-1}}$. However, many of the smaller variations in suspended sediment content do not obviously match variations in discharge. At some glaciers, the peak sediment concentration may slightly precede the maximum discharge, probably reflecting the pick-up of sediment during the phase of rising discharge. In other cases, the peak discharge may coincide with a minimum of suspended sediment if the glacier bed is relatively clean and has insufficient loose material to be incorporated into the stream.

The bedload in a stream comprises material that is rolled or bounced ("saltated") along the bed. Hydroelectric power schemes need to trap such debris, using grids over which the larger material rolls, but which allow the water with suspended sediment material through. Norwegian and Swiss engineers have gradually improved the efficiency of these traps, but nevertheless there is still considerable wear on tunnels and turbine blades from the suspended sediment.

The relative proportions of bedload and suspended sediment are highly variable; bedload is greater in steeper outwash areas than in gently inclined ones. Overall, it is common each year for the bulk of sediment discharge to take place during a few days of high-magnitude flood events.

The rôle of meltwater as an agent of chemical solution of the bedrock has often been underestimated. Meltwater with dissolved carbon dioxide forms a weak acid capable of dissolving limestone or dolomite bedrock. Selective mineral solution may also occur, for example, of feldspars in igneous and metamorphic rocks, or the breakdown of micas into clay minerals, so weakening the rock as a whole. Mineral-rich waters may leave thin deposits of calcite or silica during regelation, especially on the lee side of small bedrock bumps in front of receding glaciers (Hallet 1976). Solutional furrows and extensive films of precipitated calcite indicate that an almost continuous film of meltwater exists beneath such glaciers.

153

5.2.3 Deposition of glaciofluvial sediments

The deposition of meltwater-transported sediment takes place in a variety of settings:
  - *supraglacially*, where the finer fraction of rockfall-derived material accumulates in crevasses, pools and streams on the glacier surface;
  - *englacially*, where surface debris is washed down moulins into the internal drainage system of a glacier;
  - *subglacially*, either in tunnels cut into the base of the glacier by basal meltwater, or in channels cut into subglacial till or solid bedrock beneath the ice;
  - *ice-marginally*, where streams of both glacial and non-glacial origin are forced to flow along the ice margins where the glacier has a convex-up cross profile;
  - *proglacially*, first, in an ice-frontal position, especially in the zone between the snout and the higher ground of an end-moraine complex and, secondly, down-valley as the main meltwater stream spreads across the valley floor or plain.

Of the above, the first two are unimportant in terms of the final sedimentary product. Subglacial, ice-marginal and proximal proglacial sedimentary processes give rise to a variety of landforms (§5.3) that are frequently modified or destroyed soon after deposition, but which nevertheless are sometimes preserved in the geological record. By far the most important setting for the development and preservation of glacial meltwater sedimentary sequences is down-valley of the glacier, away from the immediate influence of marginal ice processes, as on braided outwash plains.

5.3 Glaciofluvial landforms

As noted earlier in this chapter, glacial meltwater is capable of both transporting large quantities of sediment and creating a wide variety of depositional landforms, and of eroding bedrock or unconsolidated sediment to a considerable depth. These depositional and erosional processes take place subglacially, ice-marginally and proglacially, and the resulting landforms may be classified accordingly (Table 5.1). However, the forms described represent a continuum of features, some of which are the result of both erosion and deposition.

5.3.1 Landforms resulting from subglacial meltwater erosion

The most important features resulting from subglacial meltwater erosion are a range of channels orientated approximately parallel to the former ice-flow direction, and incised into the sub-ice bed, whether bedrock or unconsolidated sediment such as till or fluvial material. Some of these forms develop as a result of a combination of ice-abrasional and meltwater-erosional processes, and are called p-forms. Where abrasion is dominant they are described in Chapter 3.

It is known from studies of the hydrology of temperate glaciers that water tends to gather into discrete streams, especially towards the topographically lowest part of a glacier cross-

Table 5.1 Classification of glaciofluvial erosional and depositional landforms according to position in relation to the glacier margin. The scale range bars refer to the maximum linear dimension of the landforms.

Position	Process	Landform	Scale (0.1 mm – 1000 km)
Subglacial	Erosion by subglacial water	Tunnel valley	~10 km – 100 km
		Subglacial gorge	~100 m – 10 km
		Nye (bedrock) channel	~10 m – 1 km
		Channel in unconsolidated sediment	~100 m – 10 km
		Glacial meltwater chute	~10 m – 100 m
		Glacial meltwater pothole	~1 m – 100 m
		Sichelwannen	~1 m – 10 m
	Deposition in subglacial channels, etc.	Esker	~100 m – 100 km
		Nye channel fill	~1 m – 100 m
		Moulin kame	~10 m – 100 m
		Carbonate film and cornices	~0.1 mm – 1 cm
Ice marginal	Ice-marginal steam erosion	Meltwater (or hillside) channel	~1 km – 100 km
	Ice contact deposition from meltwater and/or in lakes	Kame field	~1 km – 10 km
		Kame plateau	~1 km – 10 km
		Kame terrace	~100 m – 10 km
		Kame delta (delta moraine)	~100 m – 10 km
		Crevasse fillings	~10 m
Proglacial	Meltwater erosion	Scabland topography	~100 km
	Meltwater deposition	Outwash plain (sandur)	~1 km – 100 km
		Valley train	~1 km – 100 km
		Outwash fan	~1 km – 10 km
		Pitted plain	~1 km – 10 km
		Outwash delta complex	~100 m – 10 km
		Kettle hole/pond	~10 m – 10 km

sectional profile. In narrow valley glaciers, there may be just one major stream running down the ice-flow centreline, where stresses in the ice tend to be least, but in broad valley glaciers, or lobes of ice of the piedmont type, there may be a series of channels, interlinked in a complex pattern similar to braided channels, although the floors of the channels may be discordant with one another.

The largest features resulting from subglacial meltwater erosion are **tunnel valleys**. Several examples associated with the last ice sheet over northern Europe have been described (Ehlers et al. 1991). In East Anglia in England, steep-sided, deep valleys have been cut in chalk and associated bedrock, and are normally filled with glacial meltwater sands, gravels or finer material; some till may also be present. Often, tunnel valleys show a reverse gradient, for example, the so-called **Rinnen** of Denmark, northern Germany and Poland. Subglacial streams under high pressure are responsible for these features. Tunnel-valley fills associated with a late Pleistocene grounded tidewater ice margin in the Irish Sea Basin have been documented, both in coastal exposures near Dublin and in seismic profiles offshore (Eyles & McCabe 1989). The channels are steep-sided, are stacked one on top of another, and measure 10 m in depth and up to 2 km in width.

Other spectacular channels are **subglacial gorges** cut into solid bedrock. Many such channels are extremely narrow in relation to their depth, several metres compared with tens of metres.

155

They tend to be deepest and narrowest where the main valley narrows or steepens, and they cut through riegels and valley-steps in a manner that suggests the stream was trying to smooth out the longitudinal gradient.

The cross-sectional shape of a subglacial gorge is generally irregular, especially where the downstream gradient is high; but, in gently inclined reaches, a flat bottom with vertical sides is more typical. Since gorges commonly exploit weaknesses in bedrock, such as faults or dykes, they are often straight.

Gorges are cut down by subglacial meltwater under high pressure, with cavitation playing an important rôle. They may also be moulded by ice, as indicated by striations, which fills the void left as the subglacial stream cuts downwards. A good example of a still partially ice-filled gorge is that at the snout of the present-day Unterer Grindelwaldgletscher in Switzerland. Many other subglacial gorges occur throughout the Alps, Scandinavia and, on a smaller scale, the British Isles. Fine gorges tens of metres deep in Britain are to be found near Loch Broom in northwest Scotland (the Corrieshalloch Gorge) and in Glencoe; gorges occur in many other parts of Scotland, the Lake District and North Wales.

In some areas, channels may be many kilometres long, but lack a consistent downstream gradient. Instead, they continue over rises in the bedrock, even those having an amplitude of more than 100 m, the water having been able to flow uphill because it was under high pressure. Following recession of the ice, **ribbon lakes** may be left filling the lower parts of the channel system. Channels may also be cut in unconsolidated till or other forms of glacial drift. They, too, are frequently deep in relation to their width, and near-vertical sides are common. Typically they attain depths of several metres.

Figure 5.2 Nye-channel of late Pleistocene age cutting into diamictite of Neoproterozoic age, Garvellachs, Scotland.

Networks of subglacial channels, cut into both bedrock and drift, have been described from many parts of Europe and North America. In Britain, the Southern Uplands and Pennines are well endowed with such features.

On a scale an order of magnitude smaller, there are **Nye channels**. These have vertical or undercut sides, are commonly sinuous (even meandering) and contain small pools (Fig. 5.2). Nye channels are especially common in carbonate bedrock, and here chemical erosion may be largely responsible for their formation. An exceptionally fine set of Nye channels have been described from the Glacier de Tsanfleuron in Switzerland (Sharp et al. 1989), and Neo-proterozoic examples have been documented from Greenland (Moncrieff & Hambrey 1988).

Channelized meltwater-flow tends to be associated with relatively low sliding rates. Subglacial channels of all sizes, once initiated, tend to be self-perpetuating and may carry most of the water, except in times of flood. In contrast, subglacial tunnels cut in the ice and referred to as **Röthlisberger channels** are ephemeral as the creep of ice is constantly trying to close them, and often does so each winter. By the same token, creep of ice may encroach on bedrock channels at times of low discharge, allowing striations to develop, but subsequent meltwater activity tends to re-open them.

Sometimes, channels run directly down steep rock slopes. These are **meltwater chutes**, which represent the places where marginal meltstreams descend to a lower position in the glacier.

Apart from channels, subglacial meltwater may create a variety of small landforms in bedrock. Potholes are circular shafts, ranging from a few centimetres across to over 10 m in diameter and up to 20 m deep, cut into bedrock in either the valley floor or side (Fig. 5.3). Internally, the shaft reveals a spiral structure and the bottom often has a collection of well rounded boulders and cobbles. Potholes are best developed in hard crystalline rocks, and fine examples occur in Scandinavia and the Alps, both on valley floors and along valley sides. In a sense, potholes represent a plunge-pool, strongly constrained by the ice, developing especially beneath a moulin that extends to the base of a glacier. In some circumstances, potholes are created by the cavitation process associated with water under high pressure. Bowls are a less well developed form of pothole.

Of the various types of p-form, only **Sichelwannen** (German for "sickle-shaped troughs"), which are crescent-shaped depressions and scallop-like features on a hard bedrock surface

Figure 5.3 Large potholes bored into gneiss in the valley wall below Franz Josef Glacier, New Zealand. The height of this feature is about 3 m.

Figure 5.4 Bedrock scallop (Sichelwanne) formed by meltwater erosion combined with glacial abrasion on bedrock near Columbia Glacier, Alaska.

(Fig. 5.4), have a predominantly meltwater origin, with cavitation playing an important rôle. Sichelwannen may not be symmetrical, but their axis always coincides with the direction of ice-flow with the horns pointing forward. They range in length from a metre to 10m in length.

5.3.2 Landforms resulting from subglacial meltwater deposition

Channels and cavities in and below ice masses provide ready sites for the deposition of glaciofluvial material. The range of landforms formed subglacially passes imperceptibly into subaerial ice-contact forms, and as a result the use of particular terms has not been consistent. In general, the term **esker** can be applied to most subglacial linear features (although some form ice-marginally) whereas the terms **kame** and **kame terrace** (§5.3.4) refer to ice-contact deposits formed adjacent to the snout and along the flanks of a glacier, respectively. Terminological difficulties and many examples of such landforms have been thoroughly treated by Price (1973) and Embleton & King (1975).

The term esker is of Irish origin and, nowadays, is applied to elongated ridges formed subglacially, and commonly normal to a glacier snout. Eskers consist of stratified glaciofluvial deposits, especially sand and gravel (Fig. 5.5). Eskers may be sinuous or straight; they can run up hill, and sometimes they bifurcate or are beaded. When first formed, they are steep-sided, characteristically having a flat or slightly arched top. Some of the largest eskers occur in central Sweden and Finland, where they extend for a few hundred kilometres. They rarely exceed 700m in width and 50m in height and, commonly, are an order of magnitude smaller than this. The internal structure of an esker frequently comprises sand and gravel with arched

Figure 5.5 The proglacial area of the partly tidewater glacier named Comfortlessbreen, Spitsbergen. In the foreground is a well defined sand and gravel esker, forming a ridge up to 4 m high. Another esker, largely reduced to a low sinuous bank, runs approximately parallel to the ice cliff in the background. To the right is an abandoned outwash fan, formed from a subglacial meltwater stream when ice was in contact with the gravel plain and discharge was at a higher topographic level than at present (cf. frontispiece).

bedding, dipping out from the centre. Graded bedding (both normal and reverse), cross-bedding, climbing ripples, load structures, slump folds, faults and parallel lamination in clays may be other sedimentary structures present (Banerjee & McDonald 1975). The coarser material is well rounded and sorted, and clast–fabric studies have indicated a strong preferred orientation parallel to the ridge and dipping down stream. Palaeocurrents are broadly parallel to the esker trend. Eskers are generally formed subglacially by deposition from meltwater streams, sometimes by the total or partial blocking of subglacial Röthlisberger channels as discharge declines in late summer, as a result of which debris is not easily carried through the system. Eskers develop best where ice is active, but nevertheless receding, whence they can be seen to extend beyond the glacier snout. Eskers also form under stagnant ice, but then they lack continuity. Some eskers may form by deposition from sub- or englacial streams as they emerge from a receding glacier into ponded water, and pass into kames.

Since eskers are associated most commonly with active ice, they tend not to be very stable features. They are further subjected to meltwater erosion as the glacier recedes. Some areas of Finland and Sweden have eskers in abundance. Here, lakes lie between the esker networks, and the ridges themselves have been used effectively as road routes through otherwise difficult terrain. Reports of eskers in the older geological record are rare, but the Ordovician sequence in the Sahara records a number of well preserved examples (Beuf et al. 1971).

In addition to eskers formed in Röthlisberger channels cut in the ice, sediment may also fill Nye channels cut into unconsolidated drift or bedrock beneath the ice. In this way, it is possible to find trough cross-bedded sand and gravel bodies enclosed by diamict or bedrock.

Rather rare landforms, **moulin kames**, are associated with debris accumulating at the bottom of moulins and are mounds several metres high or more. They are rather ephemeral features.

5.3.3 Ice-marginal landforms resulting from stream erosion

Ice-marginal landforms resulting from stream erosion are represented by meltwater channels perched high up on valley sides and trending in the same direction as the former ice-flow. As at the bed of a glacier, ice-marginal meltwater streams may erode both the ice and the adjacent sediment or bedrock. However, as noted in §5.2.2 the two types of channel form under different glacier thermal regimes. Whereas subglacial channels are predominantly associated with temperate glaciers, **ice-marginal channels** tend to occur along the flanks of cold glaciers, since the subzero ice temperature prevents downward movement of meltwater through the glacier. Ice-marginal channels are common in the Queen Elizabeth Islands of Arctic Canada (Fig. 5.6), Greenland and Svalbard. Distinguishing between the different types of channel facilitates the refinement of palaeoclimatic reconstructions. Marginal channels are often in direct contact with the ice (Fig. 2.11), or they lie between an ice-cored lateral moraine and the valley side, or both. It is thus not uncommon to find channels running, essentially, across a hillside after a glacier has receded. Some hillsides might show a succession of meltwater channels at different levels, perhaps intimately associated with lateral moraines and having a form that mimics successively lower ice levels (Fig. 5.6). Some channels may terminate abruptly if the water has found a direct route or **chute** straight down the hillside beneath the glacier.

Overflow channels, as their name suggests, are cut by marginal streams overflowing low cols at or below ice-surface level. The water in such cases is often dammed as lakes before incision. Many such channels of Pleistocene age have been identified throughout glaciated temperate regions, but at a time before the characteristics of subglacial drainage were known. Thus, re-examination of these Pleistocene channels indicates that many are not overflow channels at all, but subglacial ones.

5.3.4 Depositional features derived from meltwater in contact with the ice margin

The area immediately in front and along the side of a glacier is sometimes characterized by a wide range of landforms, deposited by meltwater, often in proglacial lakes. These forms are **kames** and related features. They are frequently associated with eskers, and the range of transitional features has led to confusion concerning their definition. Broadly, the nomenclature of Embleton & King (1975) and Sugden & John (1976) is followed here.

The term *kame* is of Scottish origin and it refers to a broad group of flattish-topped, ice-contact glaciofluvial landforms. They have no specific orientation with respect to the ice-flow direction. They occur singly, as isolated hummocks, as broader flat plateau areas (**kame plateaux**), or as broken terraces, usually in a proglacial setting. Where a large area is covered with many discrete kames, the term **kame field** is sometimes used. Kames consist of well

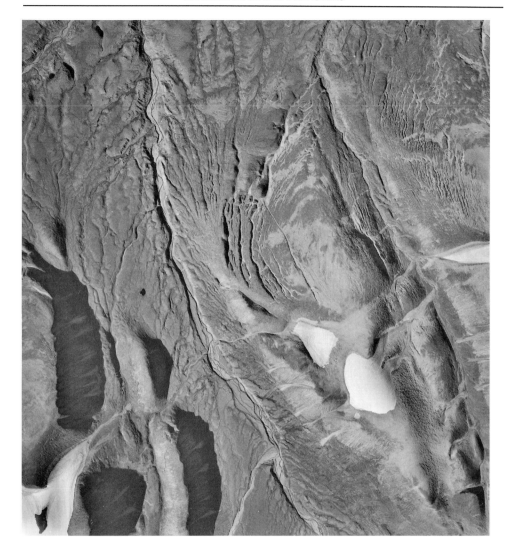

Figure 5.6 Vertical aerial photograph illustrating abandoned ice-marginal meltwater channels, SW Ellesmere Island, Northwest Territories, Canada. The channels mark successive positions of the glacier tongue as the ice receded southwards up the main valley (photograph NAPL A16780–69, with permission of the Department of Energy, Mines and Resources, Canada).

sorted, stratified sand and gravel. They vary from a few hundred metres to over a kilometre in length, and are an order of magnitude less than this in width. They form in direct contact with the ice, so when the glacier recedes, their up-valley faces are prone to slumping and collapse. Kames were formed in abundance at the margins of the last great ice sheets, especially in Canada and Scandinavia. Their development was favoured particularly by the huge volumes of meltwater released during the rapid recession of the ice margins. Embleton & King (1975) have described examples from Scotland that were associated with decaying Southern Uplands ice.

Kame terraces are distinct from kames in that they form parallel to the ice-flow direction from streams running along the flanks of a stable or slowly receding ice margin (Fig. 5.7). Their composition, however, is similar to that of kames, comprising coarse, bedded gravel and sand with planar bedding or cross-stratification if deposited by streams, or fine-grained parallel-laminated clay, silt and fine sand if deposited in an ice-marginal lake. Kame terraces slope down-valley at an angle approximating that of the former ice level. They may also tilt gently up towards the adjacent hillside, especially if the material is derived from a side valley. As the ice recedes, kame terraces are prone to collapse. In favourable circumstances, ice still-stands allow a succession of terraces to form at different levels.

Figure 5.7 Kame terrace in Adams Inlet, Alaska formed during the recession of the Glacier Bay glacier system over the past hundred years. Note that the former, steep ice-contact slope is now undergoing slumping.

Related to kames, but occurring on a much larger scale, are **delta moraines** (also known as **kame deltas**). These features form transverse to the direction of ice-flow in a marginal position where a series of subglacial streams enter a proglacial lake or the sea. Delta moraines may thus be intimately associated with eskers. In addition to typical glaciofluvial material, they also contain glacial debris derived from the ice itself. For example, ablation debris on the surface of the ice may fall directly on to the delta top. Delta moraines require the ice position to have remained relatively stable for a long period. Probably the largest delta-moraine complexes are the three Salpausselkä moraines of Finland, associated with a lake over the southern Baltic Sea region that was dammed by the Fennoscandian ice sheet (see Ch. 6).

Where glaciofluvial processes substantially rework the deposits of an end-moraine complex, the resulting feature is a **kame moraine**. The process is especially pronounced between lobes of the glacier and, here, bedded gravel and sand accumulates. The orientation of individual kame sediment accumulations is random within the context of an end-moraine system.

A last group of minor landforms are **crevasse-fillings**. They are composed of stratified drift which entered the crevasses via supraglacial meltstreams. They are linear features, a few tens of metres long, but may be of any orientation, depending on the type of crevasses that were filled.

5.3.5 Landforms resulting from meltwater erosion in a proglacial setting

Streams emerging from the snout of a glacier may be highly erosive, although less so than where they are flowing subglacially. Their typical high debris load and high velocity lead to channel-switching, and constant erosion and redeposition of proglacial sediment. Sometimes water collects in proglacial or ice-marginal lakes which, as the melt-season progresses, tend to overflow through **spillways**. The outlet streams, especially if the lake is held back only by unconsolidated sediment, may rapidly cause down-cutting and lowering of the water level. Nowadays, large proglacial lakes are rare, but they seem to have been abundant near the southern limits of the great Pleistocene ice sheets. A good example of a spillway is the now-abandoned major channel that formed when glacial Lake Agassiz, lying to the northwest of the Great Lakes, overflowed southwards into the Minnesota River and eventually into the Mississippi.

Proglacial drainage patterns in front of the former ice sheets responded to fluctuations of the ice margin, and major drainage diversions and the creation of new channels have been the result. In Britain, the best known example was the diversion of the River Thames during the Anglian glaciation from a route from the Chilterns to East Anglia, to its present course through where London now stands.

5.3.6 Depositional landforms derived from meltwater in a proglacial setting

The landforms resulting from meltwater after it leaves a glacier are not directly influenced by ice, except where stranded ice has been buried. However, in terms of volume, the amount of sediment may greatly exceed that directly deposited by the ice, at least in areas with temperate glaciers.

Braided-river systems, known as **outwash plains** or **sandar** (from the Icelandic; singular *sandur*), develop down-valley of glacier snouts that terminate on land. In the strictly Icelandic sense, sandar are laterally unconstrained, and in their type area at the southern margin of the ice caps in southeastern Iceland their width may be as great as or greater than their length. They are characterized by many active braids across most of the plain, but during jökulhlaups the whole area may be flooded.

In mountain areas, a braided-river system or **valley-train** may extend across the whole

Figure 5.8 Valley train below Mount Cook (centre-right) extending southeastwards from debris-covered Tasman Glacier (right) and Müller Glacier (centre). Note the partial superimposition of the Müller braided-river system on the Tasman system as it enters the main valley.

Figure 5.9 Kettle hole, actively forming in the outwash plain below Casement Glacier, Alaska.

width of a valley, mountains rising sharply from the valley floor. Good examples occur in Alaska and the Southern Alps of New Zealand (Fig. 5.8). In areas of postglacial uplift, successively higher levels of a braided-river system are abandoned on terraces as the river cuts downwards, the most recent active outwash plain being narrower than prior to uplift. In relation to a receding glacier, a braided system may occur within the terminal moraine, often in association with proglacial lakes. A single channel may emerge from the moraine, and a new braided-stream system may evolve outside the moraine. Glaciofluvial activity may rework much of the till and cover it almost completely. **Braided outwash fans** develop where river systems, constrained by valleys, debouch on to lowlands beyond mountain ranges. Fans often coalesce, forming a broad succession along a mountain front. Many examples may be cited, but the best known lie to the north of the European Alps.

A common characteristic of braided-river plains is the presence of water-filled pits called **kettles** or **kettle holes** (Fig. 5.9). Often, these are steep-sided and shaped like a cone when freshly formed. They are the result of the burial of dead glacier ice, either as a remnant of a glacier left behind during recession or as a detached block of ice swept down stream during a flood, in both cases being subject to burial by glaciofluvial material. Slow thaw of the ice undermines the gravel above, causing it to cave in and gather water. If an outwash plain is littered with many kettle holes, the term **pitted plain** is often used.

Braided streams generally develop in response to marked fluctuations in discharge, and as such are not confined to glacial environments; for example, ephemeral streams in deserts are

Figure 5.10 Longitudinal bars with diagonal flow in the braided river below Fox Glacier, New Zealand (cf. Fig. 5.11). Note the generally coarse, gravelly material of the upstream ends of the bars. A point bar is also present, adjacent to the rock wall in the background.

commonly braided (for comprehensive reviews of braided-river environments see Miall 1977, 1983). In glacial environments, braided streams form mainly in response to seasonal variations of flow from glaciers. In winter, discharge from a glacier is frequently reduced to a mere trickle, but in early summer, when ice melt is combined with snow melt, the entire valley floor may be washed by the flood, and much debris is transported. This represents an annual switch from empty channels to bankfull discharge. As a result, distinct sedimentary structures (bedforms) develop. The reader is referred to any modern textbook of sedimentology for an account of these forms.

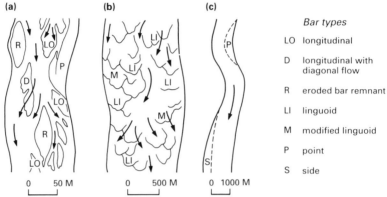

Figure 5.11 Principal types of bar in braided-river channels (from Miall 1977, with permission of Elsevier Science Publishers, Amsterdam).

Figure 5.12 Upstream-dipping cobbles and boulders (imbrication) in the main channel below Franz Josef Glacier, New Zealand.

In coarse sediments, the development of primary structures is less well known. Large dune-like features of gravel have been left behind after catastrophic lake drainage. Pebble- or boulder-ridges, transverse to flow in outwash areas, are also known.

On a larger scale, all sediment types give rise to the **channel bars** that are so typical of braided glacial outwash streams (Figs 5.10 & 11). Bars may be relatively stable and may only shift during floods or when an entire channel migrates.

In general, the range of sedimentary structures reflects the distance from the glacier, permitting us to distinguish proximal environments comprising featureless or crudely parallel-bedded or discontinuously bedded gravels, from distal environments with a wide variety of structures, including cross-bedding in sand bars, and dunes and ripples as channel bedforms. In the proximal environment, there is a wide range of grain sizes as a result of rapid deposition, except in certain backwater locations. Clast imbrication is common (Fig. 5.12) and disc-shaped clasts are tilted up stream by up to 30°. Palaeocurrent directions give a good approximation to the mean trend of channels and bars (Rust 1975). Melting of associated ice, particularly if

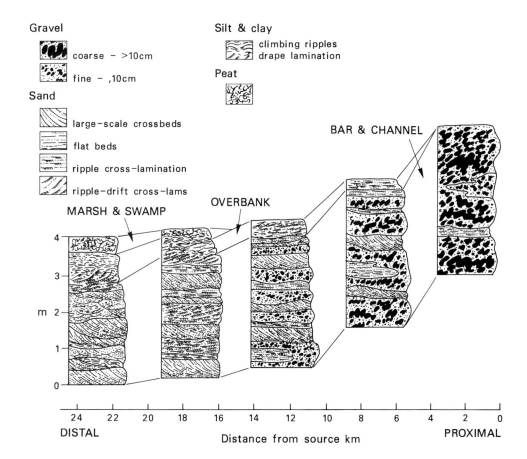

Figure 5.13 Downstream variation in facies and sedimentary structures. Bar and channel sequences do not fine upwards, but are capped by a finer overbank facies that becomes more important in a down-fan direction (adapted from Boothroyd & Ashley 1975).

167

it has been buried, will produce various high-angle faults and soft-sediment deformation structures (McDonald & Shilts 1975), as well as generating a pitted surface on the plain.

In more distal areas, foreset bedding develops on the distal, prograding ends of some bars where they build out into deeper water. Festoon structures, parallel and oblique foreset-lamination, and trough-fill structures are also common. The sediments fine upwards and downstream on linguoid bars if the depositional history is simple.

Periodic flooding may distribute material over a wide area and lead to channel migration. Deposits associated with floods are (a) channel fills, (b) **levées,** which are best developed in sandy river beds, and (c) surface veneers that comprise fine gravel to sand, the latter often being deflated to leave a lag-gravel, or having a rippled top.

The texture of braided-river deposits changes markedly down stream (Fig. 5.13). The size of material decreases down stream, but this is more noticeable in the channel beds than on the surface of the outwash plain generally. The increase in roundness initially is very rapid. Sorting by lithology also occurs, resistant rocks becoming more prominent down stream. There tends to be little lateral mixing of sediments. Stone orientation is normally weakly parallel to flow, but sometimes it is transverse. In fine-grained sediments, several characteristic grain-size distributions have been recorded, e.g. a log–normal distribution, a distribution truncated at the fine or the coarse ends of the spectrum, or a bimodal distribution.

Braided-river deposits may accumulate to depths of a few hundred metres in favourable locations, and little evidence of direct glacial deposition may survive.

5.4 Glaciofluvial facies and facies associations

5.4.1 Lithofacies and their interpretation

The lithofacies of braided glacial river systems have been classified and presented as a generalized scheme by Miall (1983b), based on the Scott River in Alaska, the Donjek and Slims rivers in the Yukon, and the non-glacial Platte River in Colorado. Lithofacies are subdivided into three main groups: gravel, sand and fines, and are qualified by terms for internal textures and sedimentary structures (Table 5.2).

Lithofacies in eskers are predominantly gravel and sand, the fines having been removed by the turbulent flow that characterizes subglacial streams. These facies become disturbed as the ice is removed. A good example of the range of facies in an esker was described by Sanderson (1975).

Kames and kame terraces, comprising glaciofluvial material, include gravel, sand and mud, with similar sedimentary structures to those in braided rivers. However, like eskers, they also show evidence of deformation or collapse following removal of the supporting ice mass.

Table 5.2 Lithofacies types in the braided-river depositional environment (summarized from Miall 1978 with lithofacies codes omitted).

Facies	Sedimentary strucutres	Interpretation
Gravel: massive, matrix-supported	None	Debris-flow deposits
Gravel: massive or crudely bedded	Horizontal bedding, imbrication	Longitudinal bars, lag deposits, sieve deposits
Gravel: stratified	Trough cross-bedding	Minor channel fills
Gravel: stratified	Planar cross-bedding	Linguoid bars or deltaic growths from older bar-remnants
Sand: medium to very coarse; may be pebbly	Solitary or grouped trough cross-bedding	Dunes (lower flow regime)
Sand: medium to very coarse; may be pebbly	Solitary or grouped planar cross-bedding	Linguoid transverse bars, sand-waves (lower flow regime)
Sand: very fine to coarse	Ripple marks of all types	Ripples (lower flow regime)
Sand: very fine to coarse; may be pebbly	Horizontal lamination, parting or streaming lineation	Planar bed flow (lower and upper flow regimes)
Sand; fine	Low angle ($< 10°$) cross-bedding	Scour-fills, crevasse-splays, antidunes
Erosional scours with intraclasts	Crude cross-bedding	Scour-fills
Sand: fine to coarse; may be pebbly	Broad, shallow scours including cross-stratification	Scour fills
Sand	High angle, planar cross-stratification, horizontal lamination, shallow scours	Aeolian deposits
Sand, silt, mud	Fine lamination, very small ripples	Overbank or waning flood deposits
Silt, mud	Laminated to massive	Backswamp deposits
Mud	Massive with freshwater molluscs	Backswamp pond deposits
Mud, silt	Massive, desiccation cracks	Overbank or drape deposits
Silt, mud	Rootlets	Seatearth
Coal, carbonaceous mud	Plants, mud-films	Swamp deposits
Carbonate	Pedogenic features	Soil

5.4.2 Braided-river facies associations

Several types of repetitive vertical sequences may be envisaged for the braided-river environment (Miall 1977):

(a) *a flood cycle* – a superimposition of beds formed at progressively decreasing energy levels;

(b) *a cycle due to lateral accretion* – a cycle generated by side or point-bar growth is possible;

(c) *a cycle due to channel aggradation* – this cycle would represent the fill of a channel or local channel system; waning energy levels would occur during sedimentation, followed by channel abandonment as a result of stream migration (**avulsion**);

(d) *a cycle due to channel reoccupation* – an abandoned, partially filled channel may be reoccupied as a result of avulsion.

Such cycles may range from 15 cm to 60 m in thickness, lateral variations may be marked, and in a given braided-stream deposit all these cycle types may be represented. Thus, interpretation of braided-river sequences requires careful and thorough field work.

Miall (1978) recognized six main types of facies association in gravel- and sand-dominated braided rivers, three of which are applicable to glacial outwash river systems (Fig. 5.14). The Scott-type facies association is named after the Scott River outwash fan in southeastern Alaska, which was studied in detail by Boothroyd & Ashley (1975). These authors described five distinct vertical profile sequences which, with increasing distance down stream are:

(a) *upper fan* – coarse gravel facies, well imbricated down stream;

(b) *upper midfan* – also gravel but finer, with thin flat beds and large-scale trough cross-bedded sands; sand-wedge deposits are important;

(c) *lower midfan* – increasing sand in relation to gravel, and large-scale festoon cross-beds increasing in abundance;

(d) *lower fan braided facies* – planar cross-bedding (resulting from slipface migration of bars) and abundant climbing-ripple lamination (also known as ripple-drift cross-lamination), resulting from bar-surface deposition;

(e) *lower fan meandering facies* – with abundant large-scale trough cross-bedding and planar to tangential cross-bedding.

Overbank deposits are absent on the upper fan, but increase in importance down-fan. The grain size in overbank silty climbing-ripple lamination and draped lamination decreases down-fan.

The Donjek type of Miall (1978) (Figure 5.14) is the most varied facies association, containing anything from 10 to 90% gravel. Marked fining-upwards cycles on several scales are present, the thicker ones reflecting either sedimentation at different topographic levels within the channel system, or successive events of vertical aggradation followed by channel switching. The differentiation of the channel–interfluve system into distinctive topographic levels is thus a prominent feature of this type of braided-river system. Bar gravels dominate the lower, most active channels, while sand and pebbly sand occur on the higher elevations, and mud may be present in abandoned areas of the outwash plain or in interfluve ponds.

A third type of association, dominated by sand, is the Platte type. Although named after a non-glacial river, it is typical of the lower reaches of a sandur. Runoff is spread between many shallow distributaries, the topographic differentiation between channels and interfluves being less than for the Donjek type, and average channel depth and slope are smaller. Sandy bedforms dominate, especially sand waves, linguoid bars and dunes, resulting in a sequence that is largely planar cross-bedded. Minor gravel and fine, overbank deposits may be present, and cyclicity is rarely observed.

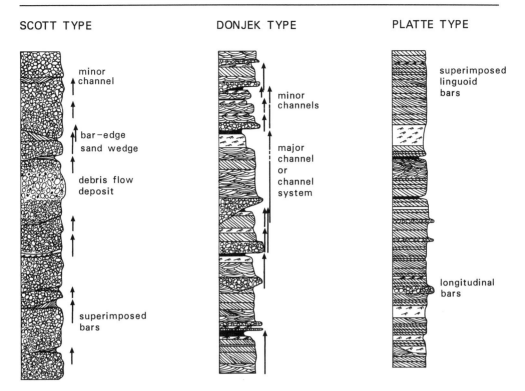

Figure 5.14 Hypothetical facies association in the most common types of glaciofluvial sequences (from Miall 1983b, with permission of Pergamon Press, Oxford).

The distal reaches of an outwash river, or where wind-blown material (loess) is abundant, may give rise to a fourth, but little studied, facies association, the Slims type, named after a river in the Yukon. Low-relief bars and channels are characteristic, and the facies are mainly massive, laminated and ripple cross-laminated sandy silts.

In a different climatic regime, namely that of the maritime Arctic of Spitsbergen, under the influence of permafrost, Bryant (1983) has provided an account of the areal distribution of facies and landforms in a proglacial braided-river valley. Comparative studies in southern Britian found many similarities with the Spitsbergen facies associations, allowing the inference that some glaciofluvial sequences were formed under periglacial conditions.

5.4.3 Facies architecture

The overall architectural styles of modern sandur or ancient outwash sequences have received little attention. Thus, the examination of environmental changes recorded by such sequences has rarely been attempted. The geometry of glaciofluvial horizons at different levels within a grounded-ice depositional sequence generally requires good sections that are both vertically and laterally extensive, for which the pre-Quaternary record offers the best prospects.

171

The architecture of a typical esker is complex. The feature itself snakes across the surface, typically a braided-river plain. Bedding rises and falls gently as observed in a longitudinal profile, while in cross section it is often gently arched and faulted. The geometry of late Ordovician esker systems has been described from the Sahara by Beuf et al. (1971). The architecture of kames and kame terraces has features which may also be found in braided-river and glaciolacustrine systems, but on a smaller scale.

5.5 Preservation potential of glaciofluvial sediments

Glacial outwash complexes commonly accumulate to depths of several hundred metres in major glacial troughs in temperate regions, such as in Alaska or New Zealand. Otherwise, they form extensive fans beyond mountain ranges over lowland areas. Many fluvial sequences have survived into the rock record, especially where deposited in fault-bounded basins; the same should be true of glaciofluvial sediments. Thus, compared with sediments released directly from ice on land, they have a markedly better preservation potential. Glaciofluvial sequences in the rock record may be more abundant than might at first be apparent, especially if direct evidence of glaciation is missing.

In contrast to braided-river complexes, eskers, kames and kame terraces are commonly subject to erosion once the glacier has receded, and only a few have survived to become part of the rock record.

6 Glaciolacustrine processes and sediments

6.1 Introduction

As discussed in Chapter 2, lakes are associated with glaciers in several different ways. Lakes in direct contact with glacier ice are referred to as glacial lakes and may form along the glacier margins, especially along its flanks, or in front of the snout (ice-marginal and proglacial lakes, respectively) (Figs 6.1 & 2). Vertical, calving ice cliffs are characteristic of glaciers that flow into lakes. Some lakes form subglacially, or from melting out of ice buried under outwash. Other lakes, that are not in direct contact with an ice margin, may still be influenced by a glacier. In such cases, a braided glacial river may connect the glacier to the lake

Figure 6.1   Ice-dammed lake at the snout of Austerdalsisen, Svartisen, Norway. Ice at the right has blocked off the low col. The lake has been artificially lowered by tunnelling, to prevent damaging jökulhlaups.

173

Figure 6.2 Proglacial lake below Mount Cook, New Zealand, resulting from recession of the debris-covered Hooker Glacier (behind iceberg) into a deeper basin. Note the suspended sediment in the water.

Figure 6.3 Saline lake in front of Canada Glacier, Victoria Land, Antarctica. Salt crystals form on the surface of the lake ice, which is a permanent feature, although because of strong thermal stratification the deeper waters may be warm. Sedimentation in such lakes is characterized by a high proportion of evaporitic minerals, and abundant stromatolites (algal mounds) occur locally.

and provide the principal sediment source; these are not strictly glacial lakes, however. Glacial lakes form in both temperate and polar climatic regimes. Some of those in Antarctica (Fig. 6.3) are highly saline and remain perpetually frozen over, yet are quite warm at depth.

Many of the largest glacial lakes of today occupy over-deepened basins dating back to the last glacial period. Such lakes are important sediment traps, and sequences a hundred or more metres thick have been preserved. The Northern Hemisphere ice sheets were bordered in many places by lakes hundreds of kilometres across, and these too have trapped much sediment.

Glacial lake processes are varied and give rise to complex assemblages of facies. These facies reflect input of material from subglacial or ice-marginal streams both as bedload, which leads to the formation of deltas, and as suspended sediment. Some sediment, carried in suspension, may settle over the entire lake floor. In addition, subaquatic gravity flow and littoral processes may operate and, in arid areas, aeolian input is also important. In many respects, the processes in glacial lakes resemble those in fjords, the latter being described in the following chapter, but there are differences due to the way in which suspended sediment interacts with the lake waters.

This chapter focuses briefly on the processes and products of glacial lakes, and emphasizes their importance during the Pleistocene Epoch. Embleton & King (1975) provide more details about specific glacial lakes, while processes and products have been thoroughly aired by Jopling & McDonald (1975), Drewry (1986) and Ashley (1989)..

6.2 Pleistocene–early Holocene glacial lakes

Glacial lakes were far more extensive at the margins of the Northern Hemisphere ice sheets than they are today, and were subject to wide and rapid fluctuations. As a result, deposits and landforms associated with them are widespread. In Europe, the largest was the Baltic ice lake which, around 10 500 years BP, stretched for some 1200 km along the southern margin of the Weichselian ice sheet over Scandinavia. An equally extensive group of lakes developed in North America as the margin of the last ice sheet there, the Wisconsinan, receded. The most important of these, inundating nearly a million square kilometres of Manitoba, Ontario, Saskatchewan, North and South Dakota, and Minnesota, has been named proglacial Lake Agassiz (Teller 1985). This lake was the result of damming of rivers flowing into Hudson Bay by glacier ice as it receded northwards. The lake came into existence about 11 700 years BP. At its greatest height, it overflowed southwards into the Minnesota River valley and on into the Mississippi River. As glacier-recession took place, outlets to the east, into the Lake Superior Basin developed. Occasional catastrophic bursts, during which up to 4000 km^3 of water were released within a year or two, created channels into Lake Superior. By 8500 years BP drainage bypassed the Great Lakes to the north, and by 7500 years BP the lake was finally drained. The widespread sediments that accumulated in Lake Agassiz included various types of varve and laminated muds, resting on late-glacial till.

175

6.3 Physical character of glacial lakes

The pattern of sedimentation in a glacial lake, to a large extent, is controlled by density differences within the water mass. Density is controlled by temperature (being greatest at $+4\,^\circ$C), the concentration of dissolved salts, and the amount of sediment in suspension. Most lakes possess a thermally controlled density stratification, which varies during the course of a summer season. Typically, in summer, a well mixed layer of low-density warmer water develops at the top of the water column, and there may be a sharp decrease in temperature at its base. In the autumn, as the surface waters cool and become denser, they sink, allowing them eventually to overturn completely. Mixing may also be induced by wind and waves.

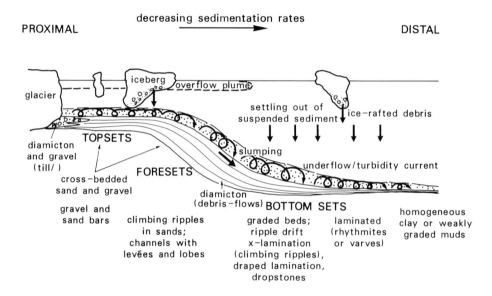

Figure 6.4 Processes and sedimentary products in a glaciolacustrine setting.

The rôle that the input of glacial meltwater has on mixing depends on the position of the glacial stream outlet with respect to water level. If there is a significant difference in density, the sediment-laden meltwater may maintain its integrity as a plume. Commonly, sediment plumes, being denser, sink to the bottom of the lake as an **underflow** (Fig. 6.4). The descending water often behaves as a turbidity current, so giving rise to graded, rhythmically stratified sediments spread over the entire lake basin floor. If a low-density (e.g. a clear) subglacial stream enters a lake containing suspended sediment, it rises to the surface and becomes an **overflow** (Fig. 6.4). Turbulent exchange between inflowing water and the lake waters may be limited unless the density contrasts are low. A third type of flow comprising water with the same density as the main body, called an **interflow**, may also develop in some lakes.

176

6.4 Glaciolacustrine sedimentary processes

Glaciolacustrine processes operate in lakes that are in direct contact with the ice margin, and (according to some workers) those that are connected to the glacier by a short river. Here we are concerned mainly with the former, in which deposition comprises the following elements (Fig. 6.4):

- direct deposition from glacier ice;
- deposition from subglacial rivers that usually enter the lake below water level;
- sedimentation from suspension;
- sedimentation from gravity-flows;
- lake-shore sedimentation;
- biogenic sedimentation;
- evaporitic mineral sedimentation.

Several processes and the resulting features are common to fjords, and are dealt with more thoroughly in Chapter 7. The most characteristic facies in a glacial lake are deltaic deposits and lake-bottom deposits (varves and other rhythmites).

In a glaciolacustrine delta situation, through which most glaciolacustrine sediment passes, deposition occurs as topsets, steeply dipping foresets and thin bottom-sets (Church & Gilbert 1975). Foresets are formed by avalanching, and to some extent by slumping. Bottom-sets are typically rhythmites, resulting from the transport of sediment by turbidity currents combined with the settling out of suspended sediment; if annual cyclicity is recognizable, they are referred to as **varves** (or **varvites** for the lithified equivalent). Varve-like sediments, linked to tidal processes, are found in fjord settings (Ch. 7), and care is needed in distinguishing between the two types. Turbidity underflow is not a continuous event, and may be significant on more than one occasion or not at all. In contrast to lakes, the salt water in fjords allows flocculation, and varves may not form. In glaciolacustrine deltas generally, redistribution of coarse-grained sediments by slumping is a significant process.

In addition to the existence of suspended matter within the water column, the formation of varves requires density stratification in the lake (Sturm 1979). Density differences are caused mainly by temperature gradients and the concentration of salts and suspended matter. Stratification may vary throughout the year. The character of laminated sediments depends, therefore, on the nature of both the input of suspended matter and the nature of stratification in the water body. Ideal clastic varves form only when there is discontinuous influx during periods when the lake is stratified.

Shoreline processes in lakes resemble those in coastal areas (see also Ch. 7), although wave action may be limited for much of the year by ice cover. Strong, down-glacier katabatic winds may cause considerable modification of beaches through wave action and shifting icebergs or lake ice onshore.

In most glacial lakes the biogenic component is low, limited by instability of the lake, the length of the freezing season, and the high sediment load, which rapidly tends to bury organisms. Some benthic organisms and plankton may survive, however, and even gather to form quite rich communities. In such cases sedimentary structures may be obliterated by burrowing animals, but trace fossils may be preserved.

Some lakes in Antarctica, notably those in the Dry Valleys of Victoria Land, show abundant signs of biochemical activity in the form of benthic microbial mats and stromatolites, mineralized to calcite, growth being possible down to a few tens of metres' depth (Parker et al. 1981, Wharton et al. 1982). Most of these lakes differ from most glacial lakes in receiving only limited meltwater, in having no outlet, and in remaining frozen at the surface all year round. These lakes have developed strong thermal stratification, and because solar radiation has been preferentially absorbed by the lower saline part of ice-covered lakes, they have become surprisingly warm at depth (e.g. 26°C in Lake Vanda). A variety of evaporitic minerals accumulate, including glauberite, halite and carbonate. These lake salts are derived from weathered bedrock and are concentrated by surface-freezing and ablation on the top surface of the lake ice, or by evaporation if the lake surface melts.

6.5 Landforms resulting from glaciolacustrine deposition

6.5.1 Deltas

Deposition of delta fronts generally takes the form of avalanching and, to a lesser extent, slumping of coarse material, against a background of sedimentation of suspended fine material (Fig. 6.2). Some delta fronts are straight, especially where meltwater channels switch from one side of a valley to the other. Where sediment is delivered from a single or narrowly confined group of channels, delta fronts are arcuate and often overlapping; they are little modified by current activity. Climbing-ripple lamination and draped lamination are characteristic of sedimentary sequences formed at delta fronts (Gustavson et al. 1975). Climbing-ripple lamination is a useful indicator of rapid sedimentation during high-discharge events.

An example of a well exposed glaciolacustrine delta, which formed earlier this century is in the Adams River Valley, Glacier Bay, Alaska (Goodwin 1984). The 200 m thick sequence comprises rhythmically bedded clays and silts with ice-rafted debris and diamictons, overlain by glacial outwash sands and gravels (Fig. 6.5). Another good example is a Weichselian glaciolacustrine delta that has been described from north Sjælland, Denmark (Clemmensen & Houmark-Nielsen 1981). Here, the foresets, which are centimetres to decimetres in thickness, are composed of coarse-grained conglomerates (matrix-supported) to sands, with slight angular unconformities. Pebbles are scattered throughout. Sedimentary structures include parallel lamination, megaripple cross-bedding, climbing-ripple lamination and isolated scour-and-fill structures. The sediments contain a fine fraction, which represents the background sedimentation from suspension.

6.5.2 Delta moraines

A **delta moraine** forms where an ice front remains stationary for a considerable time in a lake or the sea. It is the product of glaciofluvial deposition immediately in front of the ice

Figure 6.5 Glaciolacustine delta, Adams Valley, Glacier Bay, Alaska, comprising 200 m of sand and gravel. The delta was formed as a result of the post-Little Ice Age recession of the Glacier Bay complex as it dammed this side valley.

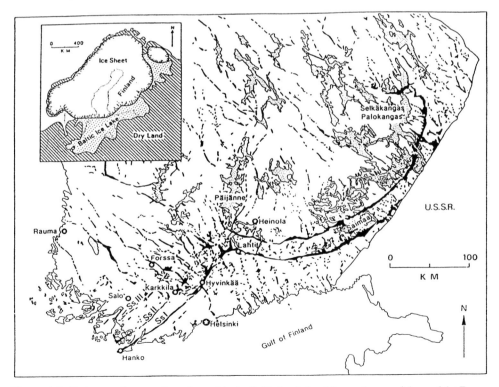

Figure 6.6 Delta-moraine complexes formed in the Baltic Ice Lake at the southern periphery of the Fennoscandian ice sheet around 10 500–10 200 years BP. The three complexes are named Salpausselkä (Ss) I, II and III. Note their association with eskers (from Fyfe 1990, with permission of Universitetsforlaget AS, Oslo).

margin. Frequently, sediment is fed into the lake subaquatically via esker channels.

Fyfe (1990) described what is probably the longest delta-moraine system exposed on land, the Salpausselkä Moraines, which extend for some 600 km across Finland (Fig. 6.6). Facies range from boulder gravels to muds, and a wide range of sedimentary structures are present. However, diamicts are uncommon. Fyfe was able to explain the marked contrasts in the form and stratigraphy along the length of the moraine in relation to variations in the water depth and the nature of the subglacial drainage system:

(a) Large individual deltas with braided tops (**ice-contact deltas**), which built up to water level, were the product of conduit-focused sedimentation.

(b) Lower, narrower coalescing fans of finer material ("grounding-line fans") were formed at the grounding-line by sediment fed from a distributed drainage system.

(c) Small, laterally overlapping subaqeous fans, derived from unstable subglacial conduit systems, occurred where marginal water depths were greatest. The differences in the subglacial conduit system are believed to be related to the reduction in the basal shear stress as the water deepens, lowering the surface ice profile and destabilizing the subglacial stream network.

6.5.3 De Geer moraines

De Geer moraines (also called washboard or sublacustrine moraines) are a group of landforms formed subglacially, transverse to ice-flow but, unlike Rogen moraines (§4.5.3), they form some way behind an ice margin that calves into the lake, especially in broad, open depressions (Sugden & John 1976). De Geer moraines are a succession of discrete, narrow ridges, ranging from short and straight to long and undulating. The ridges are more delicate than those of a Rogen moraine, although they are occasionally linked by cross-ribs. The ridges rarely exceed 15 m in height, and their spacing may be up to 300 m. They are composed of till with a cap of boulders, while lenses of sand and other stratified waterlain deposits, including varves, occur between the ridges.

The origin of these features is unclear, but may be linked to accumulation of sediment where the base of the glacier decouples from the bed of the lake, allowing debris to accumulate as a ramp between well grounded and floating ice. Each ramp ceases to develop when the ice thins sufficiently to break off as an iceberg. Another model concerning the development of De Geer moraines, based on Finnish studies, suggests that local surging phases of the Fennoscandian ice sheet may have led to flexuring of the basal ice zone, so as to create basal crevasses. Subsequent lowering of the ice mass is considered responsible for squeezing highly saturated sediment, which is then left as ridges as the ice melts (Zilliacus 1989).

6.5.4 Shorelines or strandlines

Many glacial lakes are subject to wide fluctuations, especially if ice-dammed. Hence beaches may develop at different levels, sometimes resulting in a staircase-like arrangement of shorelines or strandlines. Shorelines become evident during the draining of a lake, and some shore-lines associated with late Pleistocene ice-dammed lakes survive.

Among the best known preserved shorelines are the Parallel Roads of Lochaber in Scot-land. Glens Roy, Gloy and Spean all have glacial lake shorelines, recognized by Agassiz around 1840, and described in detail by Jamieson in 1863; they figured prominently in the presenta-tion of evidence of glaciation in Britain (Peacock & Cornish 1989).

There are three sets of "parallel roads", at 260, 325 and 350 m, each related to cols over which ice-dammed water flowed (Fig. 6.7), first of all as progressively higher cols became blocked by ice, then as it receded. These events took place during the latest glacial phase in Scotland, the Loch Lomond Stadial about 10000 years BP. The last glacial lake, at 260 m, covered an area of 73 km^2 and had a volume of 5 km^3, which is thought to have drained cata-strophically as the ice dam failed (Sissons 1979). Many of the lacustrine sediments are deformed, and some beaches have been displaced by faulting, probably as a result of an earth-quake that occurred in response to crustal adjustment as the lake emptied. Deltas that formed in the lake from side-valley streams were soon dissected by the same streams after the lake had emptied.

Figure 6.7 Glen Roy, Lochaber, Scotland showing three prominent "parallel roads" or ice-dammed lake shorelines.

181

6.6 Glaciolacustrine facies

From a knowledge of the lithology, geometry and sedimentary structures, most glaciolacustrine sediments may be interpreted as either deltaic sediments or lake-bottom sediments, and classified as topset, foreset and bottom-set beds (Fig. 6.4). The relative importance of the different facies reflects the rôle played by direct deposition from subglacial or proglacial streams, the importance of deposition from grounded or floating ice, and the fluctuations of overflow, interflow and underflow conditions.

6.6.1 Deltaic sediments

Dissection of deltaic sequences after glacial lakes have drained has often provided useful cross sections for lithofacies investigations (Fig. 6.8a). Deltaic sediments with both topsets and foresets are commonly referred to as "Gilbertian deltas", a type described in the late 19th century from Pleistocene Lake Bonneville, centred on Utah, USA. Topset beds are typically sand and gravel, and are the result of braided stream deposition (Fig. 6.8b). In a Danish Pleistocene example, Clemmensen & Houmark-Nielsen (1981) found clast-supported conglomerates, with frequent imbrication and stones up to 20 cm in diameter, forming irregular sheets, interbedded with thinner sands with horizontal lamination or rare semi-planar or trough crossbedding. Channel structures may also interfere with the upper part of the foreset beds. In a late Holocene glacial lake sedimentary complex at Austerdalsisen, Norway, Theakstone (1976) described large sets of cross-bedding up to 8 m thick, with foresets at a moderate angle (c. 20°) (Fig. 6.8a). Low-angle foresets are known from other glaciolacustrine deltas.

6.6.2 Lake-bottom sediments

On the bottom of the lake, sedimentation from turbidity currents may be discontinuous during the melt-season. Varves may be produced on the distal prodelta slope and thin away from the source, with nearly all the thinning in the coarse summer layer. Climbing-ripple lamination is common in the foreset beds of the prodelta slope. Varves contain many graded laminae and rare climbing-ripple laminations, perhaps even isolated ripples in the coarse summer layer (Gustavson et al. 1975).

In the specific case in Denmark referred to above, the bottom-set beds contain low-angle climbing ripples, draped lamination, parallel-laminated clay, sigmoidal lamination, wave-ripple lamination and disturbed lamination (Clemmensen & Houmark-Nielsen 1981). The bottom-sets of the Austerdalsisen glaciolacustrine delta (Theakstone 1976) contain evenly bedded, laminated fine silt and clay, and some fine sand. In places, the laminae are draped over obstacles. The sediments are rhythmically alternating, in part due to colour variations, but they are not true varves, and probably were deposited from suspension (Fig. 6.8 c). Deformational structures occur in laminated bottom-set sediments (Theakstone 1976), including convolute lamination, flame structures, normal faults and thrusts with displacements of a few

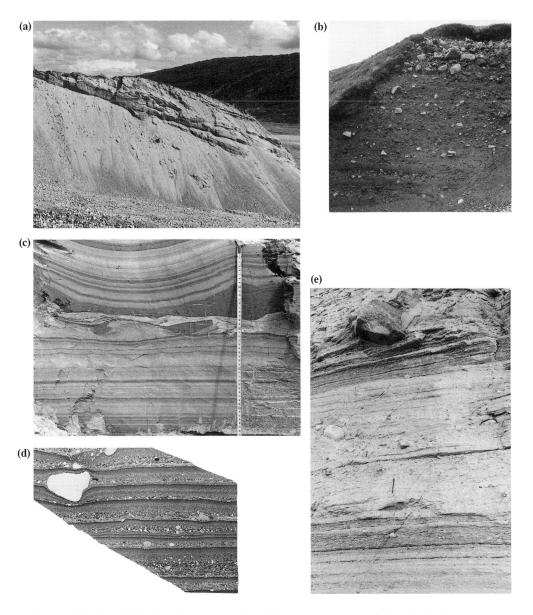

Figure 6.8 Typical lithofacies in now-exposed glaciolacustrine sequences: (a) Cross-bedded sand and gravel foresets in a 30 m section through the delta. (b) Sand and gravel topsets in a glaciolacustrine delta, upper Glen Roy, Scotland. The height of the section is about 4 m. (c) Bottom-set sand and silt rhythmites of complex origin, Austerdalsisen, Norway. The lower set is relatively undisturbed, but at the top there is a disturbed zone showing convolute lamination resulting from overpressurization of saturated sediment. The top "synclinal" structure is probably the result of slumping with the mass being detached from more stable sediment beneath. (d) Thin-section of dropstone in varvite (lithified varve) in the Neoproterozoic Elbobreen formation, NE Spitsbergen. The dropstone is about 1 cm in diameter. (e) Stratified diamicton of Pleistocene age, Taylor Valley, Victoria Land, Antarctica. During expansion of the East Antarctic ice sheet, Taylor Valley was blocked at its mouth, allowing a lake to form. This facies was derived as a result of rain-out of glacial debris from icebergs, accompanied by reworking by lake-bottom currents. Larger ice-rafted blocks have penetrated the stratification.

centimetres, and recumbent folds. Such structures are due to loading or slumping, although some may result from the melting of buried ice blocks.

Correctly identified, varves may be diagnostic of sedimentation in glacial lakes, especially if laminae are pierced by dropstones (Fig. 6.8d). Varves comprise couplets of silt and clay, representing summer and winter accumulations of sediment, respectively, and on the basis of their relative thickness they can be classified as follows (Ashley 1975):
- group I – clay thickness > silt thickness;
- group II – clay thickness = silt thickness;
- group III – clay thickness < silt thickness.

Group I varves do not occur as graded beds, but rather as two distinct layers, both the silt/clay and clay/silt contacts being fairly sharp. The silt unit is not as a whole normally graded, but consists of laminae containing minute graded beds. The clay unit, however, does show a generally decreasing grain size upwards; its upper surface may be uneven as a result of bioturbation. Clay accumulates unhindered throughout the year, but at periods of high discharge there is an influx of silt in turbidity currents. Such deposits are distal to the main sediment input source.

Group II varves are also couplets. The clay layer is graded, the silt layer shows multiple graded laminae and small-scale cross-lamination and erosional contacts. These sediments form in an intermediate position between the delta and the distal portions of the lake.

Group III varves show a considerable variation in thickness of the silt layer, whereas the clay thickness remains relatively constant, suggesting two different processes. A sharp contact occurs between the clay and silt layers. The silt layer shows laminations graded on the microscopic scale, but as a whole it does not always fine upwards. Clay layers, however, always fine upwards, suggesting that flocculation is not very important. This group is closely associated with deltas, i.e. they are formed in a proximal position. Graded beds result from turbidity currents and the erosional contacts suggest current action. Where varves or varvites contain clasts larger than the size of the layers, and reveal dropstone structures, these sediments provide some of the best evidence of glacial conditions (ice-rafting in a glaciolacustrine environment) in the rock record.

6.6.3 Rain-out diamict and subaquatic gravity flows

In lakes with a high input of ice-rafted material another facies, diamict, may be dominant (Fig. 6.8e). In a study of the lake deposits at Scarborough Bluffs, Lake Ontario, Eyles & Eyles (1983) documented the following lithofacies: diamict, sand, mud and minor gravel. This study is important because it demonstrated the importance of diamict deposition in subaquatic situations, and provided an explanation for the common gradual passage from diamict into rhythmites. The Scarborough Bluffs sequence was regarded as typical of a large, glacially influenced lake basin. Both massive and stratified diamicts are present; they consist of clayey silts with variable proportions of sand and gravel. Sedimentary structures include lenses, starved ripples, sand balls and pillows, intraformational breccias, undeformed laminated muds as interbeds, and conformable loaded contacts and diamict balls. The subaquatic origin of the diamicts was not accepted by some authors, but the sedimentary structures suggest that the diamicts formed by a variety of mechanisms, but notably by rain-out from icebergs and a floating ice margin, with fine material falling out of suspension, or by resedimentation by subaquatic gravity flows. Reworking by bottom currents led to the more gravelly concentrations. Other workers have argued that the bulk of the diamict was a lodgement till, but whichever explanation is correct, there is no doubt that in many diamict sequences, rain-out and subaquatic gravity flowage are important processes.

6.7 Glaciolacustrine facies associations

The assemblages of facies present in the glaciolacustrine environment have been described by Shaw (1975). In proximal glaciolacustrine successions (i.e. delta situations) he identified the following facies: gravel, cross-bedded sand, flat-bedded sand, cross-laminated sand, alternating sand, silt and clay, parallel-laminated sediment and diamicton. Fining upwards successions were found to result from ice-front recession. Flat-bedded sands comprise multi-storied channel deposits. Increased deposition at the distributary mouth was found to result in the formation of bars covered by rippled surfaces. At the bottom of deep-water channels, which breached distributary mouth-bars, dunes with cross-stratified cosets were present. Upward-fining trends, were sometimes interrupted as a result of channel migration. The finer grained facies were interpreted as overbank deposits.

In the Late Palaeozoic Karoo Tillite sequence of South Africa, Visser (1983a) described a vertical succession of facies that can be explained in terms of a glaciolacustrine retreat sequence. The lake bed is represented by heterogeneous diamictite, deposited during the last glacial advance over the lake floor. Next follows deformed siltstone with diamictite lenses and sandstone beds, showing evidence of debris flowage, dropstone activity and underflows, and indicating that there was floating ice on the lake. On top lies varved shale, then rhythmite, still suggesting the presence of ice, but not in contact with the lake. Finally the sequence is capped by black carbonaceous mud, indicating a marine transgression.

The Scarborough Bluffs sequence, referred to above, provides a good illustration of a glacially influenced lake-bottom association. Vertical profiles logged by Eyles & Eyles (1983) show considerable variability in detail but, broadly, a dominance of diamictons, with lesser amounts of laminated sand and mud. By following the sequence laterally, Eyles & Miall (1984) showed that three main facies associations could be traced for several kilometres: a rain-out/resedimentation diamict association, a turbidite basin association and a deltaic association. Eyles & Eyles (1983) summarized the main processes operating and the resulting facies associations in a lake basin of this type in a facies model (Fig. 6.9); this model also applies to glaciomarine settings.

Rather different facies associations are associated with arid evaporating glacial lakes in Antarctica. Modern facies investigations are limited, but an analogous Neoproterozoic succession has been described by Fairchild et al. (1989) from Spitsbergen. In addition to the more normal lake facies of rhythmites, diamictites and sandstones, there are well developed

185

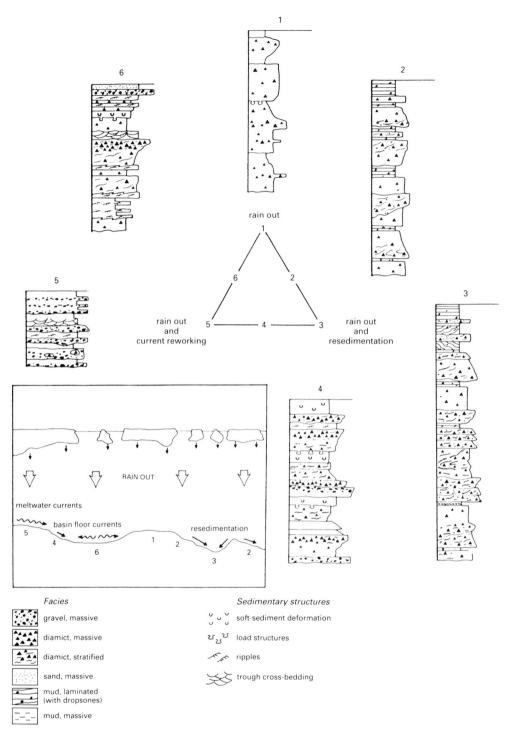

Figure 6.9 Depositional model for diamict sequences in glacial lakes (as well as massive settings) subject to substantial influxes of fine-grained suspended sediment and ice-rafted material. The principal processes and the resulting facies associations are illustrated in these idealized profiles. The numbers in the triangle refer both to the sections and the schematic diagram in the inset (adapted from Eyles & Eyles 1983).

186

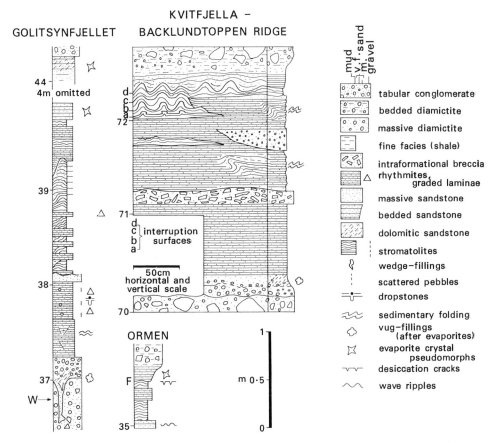

Figure 6.10 Detailed logs through a glaciolacustrine facies association of the "Antarctic type" in which biochemical processes are important; Neoproterozoic Wilsonbreen Formation, NE Spitsbergen. The Kvitfjella–Backlundtoppen Ridge section also shows the geometry of the facies as traced laterally. The vertical scales are in metres, measured from the base of the formation (from Fairchild et al. 1989, with permission of Cambridge University Press, Cambridge).

stromatolitic carbonates, which are transitional with the rhythmites. Carbonate is also a major component of the clastic facies, and evaporitic mineral pseudomorphs are common (Fig. 6.10). It is interesting to note that the common association of diamictites with carbonates and stromatolites, formerly used to discredit the Neoproterozoic glacial hypothesis, can now be viewed in the context of a modern Antarctic glaciolacustrine environment and, one might suspect, that other ancient sequences are of a similar nature.

7 Glaciomarine processes and sediments

Part 1 Fjords

7.1 Introduction to the glaciomarine environment

The glaciomarine environment is here taken to include all areas in which glaciers reach the sea and influence sedimentation, ranging from fjords (Fig. 7.1) to those areas affected by iceberg drift and far distant from the source glaciers. Until the early 1970s, our knowledge of glaciomarine environments was limited, principally because of the difficulty of undertaking systematic investigations in ice-infested waters. Nevertheless, glaciomarine sediments today are being deposited widely on contemporary continental margins, and form a substantial part of the Middle and Late Cenozoic stratigraphic record in these areas.

Figure 7.1 Grounded tidewater glacier, Comfortlessbreen, at the head of the fjord Engelskbukta, Spitsbergen. In the foreground is a sandy pebble beach, derived as a result of reworking of subglacial material.

Interest in glaciomarine environments blossomed through the 1970s and 1980s and continues to flourish today. Several volumes containing papers devoted to glaciomarine processes sediments have been published in the last 10 years (e.g. Andrews & Matsch 1983, Molnia 1983b, Dowdeswell & Scourse 1990, Anderson & Ashley 1991), as well as several review and numerous case-study papers. In addition, the textbook by Drewry (1986) covers the glaciomarine environment thoroughly, though much significant work has been published since. An advanced text on fjords by Syvitski et al. (1987) has also been published. This interest has developed mainly because we now have sophisticated research vessels equipped for effective sea-bottom sampling, coring and drilling, and undertaking detailed seismic, bathymetric and oceanographic studies. Even so, some of the most significant advances in understanding fjord environments have come about from low-budget studies from small rubber boats in Alaska. The now-considerable body of sedimentological and geophysical data from glaciomarine environments, combined with advances in the understanding of the dynamics of ice masses that reach the sea, have now given us a far better understanding of glaciomarine processes than that of 15 years ago. However, many processes remain unclarified by direct

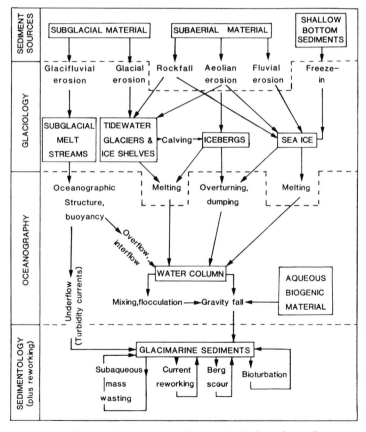

Figure 7.2 The complex system making up the glaciomarine sedimentary environment, showing sediment sources, processes and environments of deposition (from Dowdeswell 1987, with permission of the author).

189

observation and we look forward to the time when data might be obtainable directly from beneath glacier ice floating in the sea. A recent promising development has been the use of a remotely operated vehicle to investigate grounding-line processes beneath the floating tongue of Mackay Glacier, Victoria Land, Antarctica and tidewater glaciers in Alaska.

In contrast to present-day glaciomarine sequences, many ancient glaciomarine sequences are well exposed, and the geometry of such sediments may be seen much more clearly in good onshore sections. Numerous studies have been made, reflecting the common occurrence of glaciomarine sediments in all the main periods of the Earth's glacial history (Frakes 1979, Hambrey & Harland 1981, Anderson 1983, Frakes et al. 1992). Indeed, because of their high preservation potential compared with terrestrial sequences, much can be learned from the ancient glacial sediments about current processes.

The glaciomarine environment is exceedingly complex and reflects not just glacial and marine processes, but inputs from biogenic sources, and from rivers and the wind. A general scheme for all glaciomarine environments to illustrate the main sediment sources, stores, pathways and processes leading to the deposition of glacigenic material on the sea floor is given in Figure 7.2.

It is convenient to consider glaciomarine environments according to two main geographical settings:
(a) fjords (this chapter), in which sedimentation is influenced by tidewater or floating glaciers, rivers and streams, together with slope and marine processes;
(b) continental shelf and the deep ocean (Ch. 8) in which sedimentation is dominated by grounded ice margins, floating glacier tongues, ice shelves and open-marine processes.

7.2 Fjord environments with active glaciers

All fjords by definition have, at some stage in their evolution, been influenced by glaciers. Today, 25% of fjords still have active glaciers. A variety of glacier regimes may be identified.
- *Alaskan regime* – highly dynamic, grounded, temperate glaciers, characterized by rapid sedimentation that are, or have been, facilitated by rapid tectonic uplift. These glaciers are found along the coast of the Gulf of Alaska, in British Columbia and Chilean Patagonia.
- *Svalbard regime* – dynamic, grounded, slightly cold glaciers, terminating in relatively shallow fjords (< 200 m deep), in which sedimentation is influenced by large amounts of meltwater during a short summer season. This type is dominant in Svalbard, parts of the Canadian Arctic and the Soviet Arctic.
- *Greenland regime* – dynamic, floating, cold glaciers in deep fjords (> 200 m). These are typically outlet glaciers from the Greenland ice sheet or the ice caps and highland ice fields of Ellesmere Island and Baffin Island in the Canadian Arctic.
- *Antarctic maritime regime* – dynamic, cold, mainly grounded glaciers, extending to near the mouths of short fjords with limited rock exposure and restricted surface melting. They are characteristic of the northern Antarctic Peninsula.

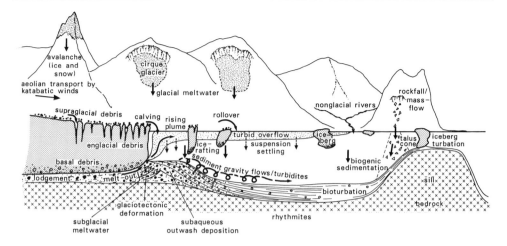

Figure 7.3 Sediment sources and processes operating in a fjord influenced by a grounded tidewater glacier.

- *Antarctic arid regime* – sluggish, very cold, floating glaciers, extending nearly to the mouth of fjords and fronted by fast ice which may survive for several seasons without breaking up; surrounded by large areas of ice-free ground dominated by aeolian processes.

Sediments in fjords are supplied from numerous sources (Fig. 7.3). Glaciers themselves provide ice-contact deposits, glaciofluvial deposits and iceberg-rafted debris. In addition, there is a marine input, including suspended sediments and biogenic material, whilst the land provides fluvial, rockfall, aeolian and gravity-flow debris. The relative importance of sedimentary inputs according to different glaciological regimes is summarized in Table 7.1.

7.3 Processes controlling fjord sedimentation

7.3.1 Deposition from glaciers

Ice-contact processes

The terminal position of a grounded tidewater glacier and floating glacier tongue fluctuates seasonally, normally advancing in winter and receding in summer. The snout is generally subject to extensional stresses and is therefore heavily crevassed. Such glaciers are therefore more dynamic than their terrestrial counterparts. Icebergs calve from vertical ice cliffs, shoot up out of the water from below the waterline, or are launched subhorizontally away from the ice cliff. Supraglacial and englacial debris falls into the water through melt-out, calving or shaking. Although prolonged observations and sampling in proximity to such ice cliffs is hazardous, a number of impressive studies have been made in recent years, notably adjacent to Alaskan glaciers, as summarized, for example, by Powell (1984), Powell & Molnia (1989), Powell (1990) and Syvitski (1989).

The rate of sedimentation from ice at the glacier terminus depends on:

- the volume of ice being melted;

191

Table 7.1 Relative importance of sedimentary inputs in different fjord regimes.

Sediment source	Debris type	Glaciological regime				
		Alaska	Svalbard	Greenland	Antarctic maritime	Antarctic arid
Direct from glacier	Subglacial debris	• •	• • •	• • •	• •	• • •
	Supraglacial debris	• •	•	•		•
	Glaciofluvial discharge	• • • • •	• • •	• • •	• • • •	•
	Supraglacial discharge	•	•			•
Marine	Icebergs	• •	•	• • •	•	•
	Sea ice		• •	•		•
	Biogenic	•	•	•	•	•
Non-direct, glacial terrestrial	Fluvial	• •	• • •	• •		•
	Mass movement	• •	•	•		•
	Wind	•	•	•		• • •

Low   High

•   • •   • • •   • • • •   • • • • •

- the debris content of the ice;
- movement of the ice front.

Sedimentation is strongly influenced by the nature of discharge from glacial streams into more dense marine waters. Submarine discharge of sediment-laden water from grounded tidewater glaciers, whether subglacial or englacial, is in the form of a jet, the behaviour of which depends on discharge (Fig. 7.4), but which invariably rises to the surface, sediment falling out as it does so. The arrival of a jet at the surface of the fjord is indicated by "boiling up" of sediment-rich waters, in which the concentration of suspended matter may be more than 50 or 60 times that of the surrounding waters.

Compared with subglacial streams, supraglacial drainage carries relatively little sediment and most is of sand grade or coarser. In many cases, supraglacial streams do not enter a fjord directly because of crevassing, but find their way to join subglacial or englacial streams. On suitable glaciers, however, supraglacial meltwater enters the fjord as a bouyant overflow, as do non-glacial or proglacial streams. Alternatively, the meltwater falls over the cliff and plunges into the stratified marine waters below. Most sediment falls directly to the sea bed at this point (Fig. 7.3).

Iceberg calving is important sedimentologically through:

(a) the release of loose supraglacial sediment – this varies according to the transverse velocity profile of the glacier and the location of supraglacial debris;

(b) generating waves – large waves may be generated several times a day, and in Alaska major calving events can devastate shorelines up to 100km from the glacier (it is not advisable to camp close to the beach where calving glaciers are active!); waves can also lift ice-rafted material above the high-tide mark;

(c) controlling the position of the ice front in combination with the forward movement of the glacier.

The contribution of the various processes to the total sediment yield has been estimated for Coronation Glacier, an outlet glacier of the Penny Ice Cap on Baffin Island, as follows (Syvitski 1989):

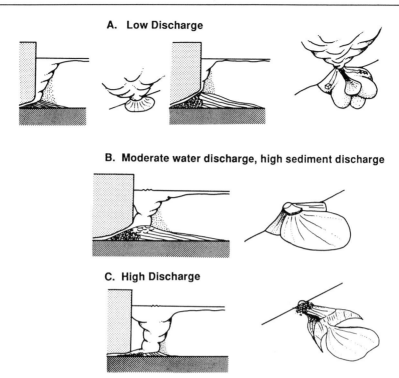

Figure 7.4   The variability of the sediment plume according to differences in discharge and the types of sediment fan associated with these differences (from Powell 1990, with permission of Geological Society Publishing House, Bath).

– 86% glaciofluvial discharge;
– 9% supraglacial dumping;
– 3.7% subglacial deposition;
– 0.8% ice-rafted englacial deposition.

The total sediment output from this glacier amounts to nearly two million tonnes a year, a figure which may be regarded as typical of many medium-sized cold glaciers. In Alaska the sediment budgets are considerably higher, but in Antarctica much less.

The style of sedimentation may change significantly if the glacier surges (Elverhøi et al. 1983) because of:

– substantially increased meltwater discharge;
– a radical change of the subglacial meltwater channel system to sheet-flood;
– erosion and reworking of glaciomarine sediments during a surge.

Many tidewater glaciers in Alaska and the Arctic are of the surging type.

Little is known concerning sedimentation at the grounding-line of floating fjord glaciers where ice loses contact with the bed. Submarine discharge of water probably occurs as a jet, and ice-contact processes may resemble those of grounded tidewater glaciers. However, the sediment plume will be prevented from rising by the floating ice, and the suspended sediment may be thoroughly mixed in the water column if, as seems likely, there is strong circulation near the grounding-line.

193

Floating ice influences

Although volumetrically unimportant, the recognition of iceberg-rafted stones or dropstones (Fig. 7.5) in stratified sediments is one of the most important criteria for establishing a glacigenic origin of a rock sequence, especially in the pre-Pleistocene record. Compared with deposition at the grounding-line, deposition from floating ice (icebergs and sea ice) is relatively minor in most fjords (Fig. 7.3). The main controls on sedimentation are:

- the concentration and distribution of debris in the source glacier;
- the residence time of an iceberg in a fjord;
- the volume of ice calved;
- the rate of iceberg drift;
- the rate of iceberg melting;
- the amount of wave action, thereby influencing the number of overturning events.

Figure 7.5   Debris-laden iceberg, a typical illustration of ice-rafting processes, Columbia Glacier, Prince William Sound, Alaska.

The dominant control on the amount of debris in the parent ice mass is its thermal regime (Dowdeswell & Murray 1990). Temperate glaciers have basal debris-rich layers of the order of centimetres to a few metres in thickness. In cold glaciers the debris-rich layers are commonly repeated by ice-tectonic processes, such as folding and thrusting, especially in the transition zone from basal sliding to a frozen bed. At the snouts of such glaciers, the debris-rich zone may be several metres thick, although sediment is unevenly distributed. Glaciers in the highest latitudes, in which meltwater is limited, entrain debris by overriding a frontal apron composed of ice and debris, creating debris layers several metres thick. Surge-type glaciers incorporate debris to such an extent that the debris layer may reach 10m in thickness; this largely arises from ice-tectonic processes, combined with flow over overpressurized basal sediments. In most cases, the debris is parallel to an ice foliation, and may account for 50% by weight of the debris-rich ice.

Estimation of sedimentation rates from icebergs in fjords is difficult, even where the total glaciomarine sedimentation rate is known. However, a two-dimensional model developed for tidewater glaciers in fjords by Dowdeswell & Murray (1990) suggests (within an order of magnitude) that the rate for Alaskan fjords of $14\,\text{mm}\,\text{yr}^{-1}$ is typical, representing 0.2–0.7% of the total glaciomarine sediment, and that in Svalbard the rate is $1\,\text{mm}\,\text{yr}^{-1}$, that is 0.1–0.3% of the total.

Sea-ice formation and break-up is an important process in the fjords of polar regions. During freezing, debris on the shore may be incorporated into the ice, later to be carried offshore under the influence of tidal currents. Material up to cobble size may be readily transported, creating a problem in discriminating iceberg-rafted from sea-ice-rafted sediments. However, sediments derived from beaches will generally be better sorted and rounded.

7.3.2 Marine processes

Water circulation

The distribution of sediment in a fjord is largely controlled by circulation in the upper part of the water column, even in deep fjords. The juxtapositon of different water masses, namely fresh clean water, fresh turbid water and sea water, gives rise to water stratification. Stratification is most marked in summer, when runoff and sediment transfer to the fjord are greatest.

A fjord often has a surface layer of fresh water derived from glacier streams, melting snow, ice and rain, most of which generally flows out of the fjord. Much of the sediment delivered to the continental shelf outside the fjord is transferred during just a few weeks, when discharge is at its peak. Sediment may only be discharged from some longer fjords every few decades. Down-fjord changes to the cross-sectional shape of the basin affect water-flow velocity; for example, constrictions cause flow to accelerate and reduce sedimentation. Surface layer velocity is affected by the tides.

Inflows of turbid freshwater plumes in areas of high precipitation and rapid ice-recession, such as Alaska, may be dense enough to flow down to, and along, the fjord floor. Evidence for these underflows comes from coarse-grained sediments within $0.5\,\text{km}$ of glaciers in Muir Inlet, Alaska. Such activity could have been important elsewhere when ice caps and glaciers were receding rapidly up fjords. These underflows can inundate an animal community on the bottom of the fjord or discourage its establishment.

The sporadic renewal of denser deep water is an important process because it prevents stagnation and oxygen-depletion of the fjord waters. When stagnation does occur, organic matter is preferentially preserved, and minerals, typical of reducing environments, such as pyrite (iron sulphide), may form. Deep-water-flow may also sweep away fine sediments on the fjord sill (if present) and redeposit them in the inner basin. Another factor is that, depending on its turbidity, deep water may either flush out a fjord (if with low turbidity) or add sediment.

Enhanced settling of sedimentary particles

Most sedimentary particles enter a fjord as single grains, and the rapidity with which they are deposited is mainly a function of size. However, enhanced particle settling may result from a number of different processes occurring at different levels in the water column (Fig. 7.6):

– *flocculation* – fine particles are attracted to each other when the normally repulsive electrostatic forces on their surfaces are neutralized by saline waters, for example, near the ice front;

– *agglomeration* – attachment of sediment grains to each other by organic matter;

– *pelletization* – the result of zooplankton ingesting sedimentary particles and egesting them as faecal pellets which then sink rapidly; suspended particles may settle as discrete layers or stringers;

– deposition of ice-welded pellets of mud or till derived from the glacier.

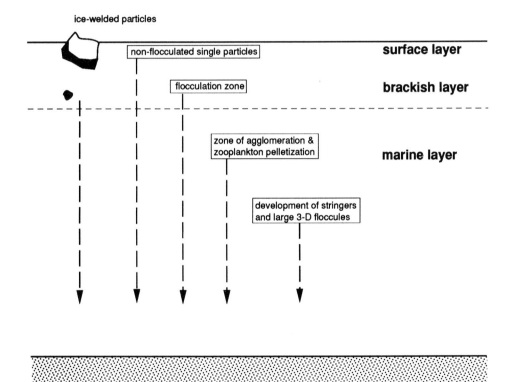

Figure 7.6 Processes responsible for enhanced particle settling in fjords (adapted from Syvitski 1989).

Tides

Reworking of sediments by tides is an important process in some fjords. Those fjords that have, at their mouth, a shallow sill may be prone to a variety of tide-related disturbances to the water. The sea-floor sediment, especially on the sill itself, may be winnowed as a result;

196

the fines are transported either seawards to the shelf or into the fjord basin, leaving a relatively coarse, sandy and gravelly lag deposit. Even where no sill exists, tidal currents may be powerful enough to rework the fjord floor, carrying out of the system fine sand and mud from the inner parts of the fjord.

Waves

Wave action is mostly dissipated at the fjord sill, but some fjords, such as those in Greenland are open to the full fetch of the ocean. Calving of icebergs can also generate large waves. Where waves have strongly influenced shorelines, nearshore zones of sand and gravel are often followed offshore by a zone of stony lag sediment and washed bedrock, and then by poorly sorted sandy to muddy sediment. In Alaska, earthquake-generated waves may cause considerable devastation (§ 7.3.3).

Submarine slides and gravity-flows

These processes are common as a result of overloading or oversteepening of slopes, especially as a result of high sedimentation rates near an ice margin. Part of a failed mass may develop into a fast-flowing turbid gravity-flow, which on slopes is confined to gullies, but beyond spreads out over the fjord basin, filling hollows in the bed, as water fills a pond.

Biogenic processes

Like many other natural water bodies, fjords provide a habitat for a variety of marine organisms, which may contribute significantly to the sedimentary pile on the fjord floor. In Spitsbergen, diatom blooms in springtime contribute to sea-floor sediments in Kongsfjorden, while in southeastern Alaska prolific, bi-annual diatom blooms are reflected in the silica content of the bottom sediment. Diatoms are also a major component of the sediments in the outermost parts of Antarctic fjords. Minor biogenic components are represented by radiolaria and silicoflagellates; rare calcareous species of planktonic foraminifera may also occur.

Larger organisms are molluscs, crustaceans and worms, which usually are sparsely disseminated through the sediment. Nevertheless, they play an important rôle in mixing up (bioturbating) the sediment and disrupting, or even destroying the stratification.

Earth's rotation

Sedimentation patterns in fjords are partly controlled by the **Coriolis force**, which is the effect of the Earth's rotation on a freely moving water mass. In the Northern Hemisphere, sediment-laden streams entering a fjord are commonly deflected towards the right-hand shore, and in the Southern Hemisphere to the left. Tongues or wedges of sediment tend to pile up along the respective sides of the fjord as a result. The effect increases in a poleward direction.

7.3.3. Influences on fjord sedimentation from land sources

In temperate and slightly cold glaciers, the dominant direct influence on fjord sedimentation from land that is ice-free near the shore, is from rivers and streams (Table 7.1). Often these rivers emanate from glaciers that terminate some distance from the sea, and so carry abundant suspended sediment. In a typical Greenland fjord, rivers are locally important, but relatively insignificant in terms of the overall length of the fjord. In maritime Antarctica, most of the ground, other than rocky cliffs, is glacier-covered, and independent streams are rare. In arid parts of Antarctica, such as the Dry Valleys region of Victoria Land, stream activity does occur, but is ephemeral, and the main influence is from wind action on till-draped hillsides.

Mass movements in the form of rockfalls, slumps or debris slides are common, although they are usually small and localized. Alaska is exceptional in this respect, because, periodically, some of the world's most powerful earthquakes generate huge rockslides. Perhaps the most dramatic event took place in Lituya Bay in July 1958, during a powerful earthquake along one of the region's major faults, the Fairweather. A landslide of 400 m³ of rain-soaked rubble was dislodged from a steep headwall, shearing off ice at the snout of the tidewater Lituya Glacier, and then riding an air cushion up to an elevation of 525 m on the other side of the fjord, leaving, in its wake, rock stripped bare of soil and forest. A huge wave raced down-fjord at a speed estimated at 250 km hr⁻¹, ripping away vegetation as it went. One fishing boat was swept out of the fjord, over the exposed moraine at the mouth, into the open ocean, in spite of which, the two people aboard the vessel managed to survive (US National Parks Service 1983). Even so, the total sediment input to these fjords from these slides represents only a relatively small proportion of the total sediment budget.

7.4 Patterns and rates of sedimentation in fjords

The patterns and rates of sedimentation are highly variable, and depend upon whether tidewater glaciers or floating tongues terminate in a fjord, or only supply debris over land through meltwater streams, as well as upon the various processes described above. In tidewater glacier environments, coarse gravel and sand tend to accumulate close to the glacier, but may occur patchily elsewhere, such as where diamictons have been winnowed by currents or where debris has fallen out of icebergs. Subglacial meltwater streams issuing from the glacier, which in many fjords provide the bulk of the sediment, deposit gravel, sand and mud, frequently draping diamicton deposited earlier. Ice-proximal mud tends to be laminated as a result of variations in discharge and tidal currents, often occurring as graded couplets. Mud extends from the glacier cliff into the distal parts of the fjord basin, and beyond if there is no sill, but ceases to be laminated because of low sedimentation rates, flocculation and bioturbation (Powell 1990).

Laminated sediments (**laminites**) have also been recorded from proximal settings in Antarctic fjords. In the maritime Antarctic Peninsula, glaciers are mainly grounded as far out as the mouth of the fjords they occupy, so relatively open bays are typical (Domack 1990). In such cases, although there are few visual signs of meltwater, deposition is inferred to be from suspended sediment in meltwater and sediment gravity-flows issuing from the ice/bed interface. However, no plumes of sediment have been observed, so a process involving a combination of basal melting, and tidal pumping and flushing of subglacial sediment is believed to take place. The direct tidal signal is masked by low sedimentation rates and bioturbation. The organic content is low in these proximal sediments, but increases away from the ice margin. In the direction of the outer bay, the biosiliceous component increases markedly, and on the continental shelf diatom ooze, with a minor amount of poorly sorted sand and gravel, is dominant.

Laminites in colder Antarctic fjords, for example, Ferrar Fiord in Victoria Land, form a very minor fraction of the sediments so far investigated. This suggests that in these areas sedimentation from subaquatic stream discharge is minimal, and that the mud fraction that occurs in both the ice-proximal and ice-distal sediments is mainly derived from basal melt-out sediment, and carried away in suspension more-or-less continuously.

An important feature of fjords is that major variations in sedimentation occur through time, both diurnally and seasonally, as well as at random, factors which contribute to the overall complexity of the sedimentation pattern. Many fjords have more than one distinct basin. In a typical fjord in Svalbard, the sediments are thickest in the innermost basin near the glacier; here over 100 m of till and compacted glacigenic sediment have accumulated. Outside the inner basin, the fjord has between 20 and 60 m of sediment, consisting principally of till or ice front or surge deposits (Elverhøi et al. 1983). The fjords in southern Alaska, described by Powell & Molnia (1989), have much greater quantities of sediment (several hundred to over a thousand metres) owing to rapid crustal uplift, a process which is responsible for one of the most active erosion systems in the world.

In both these cases the fjord sediments are young, probably dating mainly from the last glaciation. In contrast, a fjord and glacial lake sequence drilled to basement in Ferrar Fiord, Antarctica, comprises a sedimentary record extending back into the Pliocene Epoch, at least 4 Ma; however, despite the relatively long time interval represented, there is only 166 m of sediment (Barrett & Hambrey 1992).

In fjords lacking direct contact with a glacier, meltwater retains its dominant rôle, but much of the coarser material is trapped in prograding deltas or fans before, or as it enters, the fjord. Beyond these deltas, mud is the dominant sediment, derived mainly from the turbid waters of glacier-fed streams.

The rates of sedimentation in glacier-influenced fjords are exceedingly variable (Table 7.2). Glacier Bay in Alaska has yielded the most exceptional values with some $9 \, \mathrm{m \, yr^{-1}}$ in its inner part, although $0.5 \, \mathrm{m \, yr^{-1}}$ is perhaps a more typical figure close to most glacier margins in temperate and cold regimes. In the more distal reaches of a fjord sedimentation rates are an order of magnitude less. It has been demonstrated, using data from a number of different glaciological regimes (Boulton 1990), that sedimentation rates decline logarithmically with distance from the ice front.

Table 7.2 Some recorded sedimentation rates in glacier-influenced fjords.

Fjord	Relation to ice front	Sedimentation rate	Source
McBride Inlet, Glacier Bay, Alaska	Grounding line Submarine outwash fan	$>13\,\text{m}\,\text{yr}^{-1}$ $5\,\text{m}\,\text{yr}^{-1}$	Powell & Molnia 1989
Muir Inlet, Glacier Bay, Alaska	Inner fjord	$9\,\text{m}\,\text{yr}^{-1}$	Molnia 1983
Glacier Bay, Alaska	Outer fjord	$>4.4\,\text{m}\,\text{yr}^{-1}$	Powell 1981
Kongsfjorden, Svalbard	Inner fjord Central fjord	$50\text{–}100\,\text{mm}\,\text{yr}^{-1}$ $0.4\,\text{mm}\,\text{yr}^{-1}$	Elverhøi et al. 1983
Van Mijenfjorden, Svalbard	Inner fjord	$15\,\text{mm}\,\text{yr}^{-1}$	Elverhøi et al. 1983
Coronation Fjord, Baffin Island, N. W. T.	1 km 10–30 km 15 km	$400\,\text{mm}\,\text{yr}^{-1}$ $2\text{–}9\,\text{mm}\,\text{yr}^{-1}$ $3.6\,\text{mm}\,\text{yr}^{-1}$ (theoretical)	Syvitski 1989

7.5 Textural characteristics of fjord sediments

The subgravel-sized fraction of sediments has been investigated in a number of fjords. Boulton (1990) has focused on the grain-size distribution of the sediment plume from Kongsvegen in Svalbard and the resulting sediments in Kongsfjorden. A series of sediment traps indicate a skewed grain-size distribution with a prominent peak at the sand/silt boundary. As the coarser material settles out first, the peak shifts down-fjord progressively into the silt mode. Fjord-bottom sediments show the same trend. For comparison, Boulton also showed the grain-size distribution for an iceberg and the resulting sediment, which proved to be similar; in this case, the sand component was even greater than that in the early plume sediments, but there was no pronounced peak.

Elverhøi et al. (1983) have examined the grain-size distribution of other facies in Kongsfjorden. The trend from poorly sorted basal tills, through somewhat muddier glaciomarine deposits on slopes and sills, to the muds of glaciomarine origin in the distal basins is clear.

The grain-size distribution of the principal facies in the sequence of Ferrar Fjord (Barrett & Hambrey 1992), demonstrates the poorly sorted nature of the dominant facies, diamictite, although a pronounced peak at the sand/silt boundary is evident in some samples. The well-sorted nature of some stratified sands is evident, supporting the intepretation of an aeolian origin.

The roundness/sphericity characteristics of clasts within fjord sediments are similar to those in terrestrial glacigenic facies. Dowdeswell (1986) investigated the shapes of the various components associated with a fjord glacier on Baffin Island. The results were similar to those for debris in transport as recorded from terrestrial locations (cf. Fig. 1.4). There are few data on the fabric and surface markings of clasts in the gravelly sediments of fjords.

7.6   Facies analysis of fjord sediments

The characterization of facies in modern fjords has largely followed a process-orientated approach (e.g. Powell 1981, Elverhøi et al. 1983, Powell & Molnia 1989). Objective designation of lithofacies in the manner advocated by Eyles et al. (1983) has only been attempted infrequently. Much of our understanding concerning fjord facies and processes has come from the temperate glacier-fed Alaskan coast. The principal lithofacies, in so far as they can be gleaned from the literature, are listed and interpreted in Table 7.3. (Note that in this book the term "lithofacies" is used differently from that in the papers describing Alaskan fjords, but in a manner more compatible with normal sedimentological treatment (e.g. Eyles et al. 1983).) Contrasting examples from Antarctica and from the older geological record are also described below.

Table 7.3 Principal lithofacies in Alaskan-type fjords and their interpretation.

Lithofacies	Interpretation
Diamicton	(a) Lodgement till
	(b) Subglacial and meltout till
	(c) Subaqueous sediment gravity flows
	(d) Silt and clay from meltwater streams + iceberg debris
Gravel (poorly sorted)	(a) Subaqueous outwash: grounding-line fan ice-contact delta fluviodeltaic complex
	(b) Gravity flow
	(c) Lag deposit from winnowed diamicton
Gravel (well soretd)	(a) Beach
	(b) River mouth bar
Sand (poorly sorted)	(a) Subaqueous outwash
	(b) Gravity flow
Sand (well sorted)	(a) Beach
	(b) River mouth bar
Mud with dispersed clasts	Meltstream derived silt + clay, with minor ice-rafted debris ("iceberg zone mud")
Mud	(a) Meltstream derived silt + clay
	(b) Tidal flat sediment
Rhythmites (laminated sand and mud)	Deposition from underflows generated from meltwater streams, and influenced mainly by tides (cyclopels and cyclopsams)

7.6.1 Facies associations forming today in Alaskan fjords

Fjord lithofacies often occur in different associations. For example, a morainal bank (§ 7.7.1) comprises a chaotic mixture of diamicton, gravel, rubble and sand (Powell & Molnia 1989) and a grounding-line fan comprises some, or all, of these in different proportions (cf. § 7.6).

Figure 7.7 Rapidly receding McBride Glacier, Glacier Bay, Alaska. The water depth at the grounded ice cliff is over 100 m.

In a study of tidewater glaciers in Glacier Bay, Powell (1981) recognized five facies associations.

- *Association I* – facies of a rapidly receding tidewater glacier with its ice front calving in deep water (Fig. 7.7). Interpretative facies deposited close to the ice front consist of reworked subglacial till, subglacial stream gravel and sand, scattered and dumped coarse-grained supraglacial debris, and low ice-push (or De Geer) moraines from winter re-advance.
- *Association II* – facies of a slowly receding tidewater glacier with its ice front calving predominantly in shallow water. Ice-front recession has been retarded or stopped by a side-wall construction or a ramp in the fjord floor, but calving is still able to proceed apace. The resulting interpretative facies are ice-proximal, coarse-grained, morainal-bank deposits, with an ice-contact delta or grounding-line fan (§ 7.7). Subaquatic gravity-flows are common down the bank foreslope producing intertonguing sand layers within the more distally deposited iceberg zone mud (Fig. 7.8).
- *Association III* – facies deposited by a slowly receding or advancing tidewater glacier, rarely calving into shallow water. When ice fronts recede or advance into shallow water they often terminate in a protected bay and lose ice by surface-melting. Lateral streams dominate the environment close to the ice front. The resulting facies is an iceberg-zone mud, deposited both close to the ice front, away from stream outlets, and in more distal locations. Large ice-contact deltas, comprising gravel to mud facies that are structureless at stream outlets, develop along the ice cliff and further down the fjord. These pass laterally into muddy sand and interlaminated sand and mud in the prodelta area and beyond, to which is added ice-rafted debris depending on the amount of supraglacial debris available.

Figure 7.8 Ice-contact delta in front of Riggs Glacier, Glacier Bay, Alaska. At high tide this delta is under water and the glacier prone to minor calving.

– *Association IV* – facies of a turbid outwash fjord. In this case, the glacier is terrestrial and produces a large outwash delta that progrades into the fjord. The resulting facies are coarse-grained fluvial deposits on the delta surface. The slope comprises sand and gravel that intertongues with marine glacial outwash mud in more distal areas. Very little ice-rafted debris is present, since only small icebergs are introduced via the meltstreams. Sand/silt rhythmites of turbidity current origin cover the floor of the fjord.

– *Association V* – facies of a shallow-water environment, distant from the ice front. This association comprises tidal-flat muds and braided-stream deposits and beach sands. These areas may be uplifted isostatically and disturbed by stranded icebergs. Eventually, this association acquires a cover of vegetation, which may be buried during a subsequent glacial re-advance.

This classification underestimates the importance of temporal and spatial variations that result from changes in the position of the glacier terminus with respect to the depositional site. In many Alaskan fjords, where rapid recession is taking place, the complexity of the facies associations is even greater (Molnia 1983a,b).

7.6.2 Plio-Pleistocene facies associations from an Antarctic fjord

The main facies in the Ferrar Fiord succession of Antarctica are massive to well stratified diamictite, mudstone and sand, and minor rhythmite (Table 7.4) (Barrett & Hambrey 1992). This association of facies is clearly different from that in fjords where meltwater is dominant. There is little mud, such as might be associated with a meltwater plume, or gravel typical of subaquatic outwash. Sedimentation was principally the result of:

- lodgement of till as ice advanced across the fjord floor;
- release of waterlain till from a floating ice tongue near the grounding-line; and
- deposition of sand that had blown on to lake ice.

Clast provenance and Quaternary geomorphological studies indicate that the glaciers which created the fjords tended to recede during glacial maximum, when the main East Antarctic ice sheet expanded and blocked the entrance to the fjord, creating a lake. Conditions may have been cold enough for a semi-permanent ice cover to be maintained on the lake, discouraging melting, so the main sediment source, other than direct deposition from ice, was aeolian.

Table 7.4 Features and interpretation of lithofacies in the cored CIROS–2 fjord sequence, Ferrar Fiord, Victoria Land, Antarctica (Barrett & Hambrey 1992).

Facies	Sedimentary features	Interpretation
Massive diamictite	Non-stratified; clasts uniformly distributed; horizontal shear surfaces; occasional striated clasts	Lodgement till; less commonly basal meltout till or waterlain till
Weakly stratified diamictite	Weak, wispy stratification in matrix, and dispersed clasts, including dropstones	Proximal glaciomarine/glaciolacustrine sediment derived from basal debris melted out near the grounding-line; some winnowing
Well stratified diamictite	Well developed stratification in matrix and dispersed clasts, including dropstones	Proximal to distal glaciomarine/ glaciolacustrine sediment; strong bottom reworking
Massive sandstone	Fine to very fine, uniform, unstratified; a few beds are graded	Aeolian or supraglacial sand deposited directly or washed off floating ice; graded beds may be turbidities
Stratified sandstone	Fine to very fine; occasional fine horizontal mm–cm laminae	As above, but with intermittent deposition
Mudstone	Moderately and weakly stratified to unstratified; little sand and few gravel sized clasts	Sedimentation from suspension well beyond ice margin, originating from basal debris close to the grounding-line; little from sea or berg ice
Rhythmite	Alternation of well sorted, very fine sand and clayey silt layers from a few mm to a few cm thick; no dispersed gravel clasts	Sedimentation from intermittent underflows of fines winnowed from basal glacial debris with background sedimentation of aeolian sand; some may be varves, no ice-rafting

7.6.3 Submarine debris-flow facies associations from the Late Palaeozoic sequence of South Africa

The importance of subaquatic debris flowage close to the grounding-line was recognized by Visser (1983b) from studies of a well-exposed sequence of Late Palaeozoic age in the Kalahari Basin of southern Africa. Visser distinguished several types of flow, based on lithology and texture, derived from material deposited near the grounding-line of a floating glacier.

These flows occurred either as individual or stacked beds, and are described and interpreted as below.

- *Type I* – IA – diamictite, massive, medium-coarse, clast-rich individual beds about 1 m thick. The diamictite grades into clast-supported conglomerate. IB – deformed argillaceous diamictite with thin, clay-rich laminae or flow-banding. There is an upward transition to rhythmite-shale. These sediments represent cohesive debris-flows with laminar flow characteristics. IB shows features indicative of deformation of a late stage in the flow.

- *Type II* – upwards-fining conglomerate and sand beds. IIA – basal conglomerate/coarse sandstone at the base, passing upwards, first into medium- to fine-grained sandstone with mud clasts, then into an upper fine-grained unit characterized by ripple marks and water-escape structures, and finally into a mudstone/rhythmite shale. This association represents a high-density turbidity-flow, in which complete liquefaction occurred. IIB is similar to IIA but contains, in addition, ice-rafted debris and angular fragments of shale in a mudstone instead of the upper fine-grained unit. This represents a transitional type of flow between the completely liquefied and the cohesive varieties.

- *Type III* – reversly graded sandstone beds. These are transitional between cohesive debris-flows and high-density turbidity currents or, in part, grain-flow.

- *Type IV* – upwards-fining sandstone/siltstone, representing a low-density turbidity current.

A sequence that shows an association of these various types of flow deposit is shown in Figure 7.9.

7.7 Depositional and soft-sediment erosional features in fjords

The wide range of processes associated with receding glaciers in fjords gives rise to depositional assemblages and erosional features that are distinct from those deposited by glaciers and meltwater on land, but which have some similarities with those in large lakes. In most cases, continued sedimentation during ice-recession, or destruction during ice-advance will preclude their preservation as distinct bathymetric features. Only rarely have these features been uplifted and eroded to allow ready inspection of their internal sedimentary structures. Our knowledge is thus based, to a large extent, on the investigations that have been made in Alaskan and Arctic fjords, and a variety of geophysical techniques, such as seismic profiling and side-scan sonar. The forms described below are part of a continuous spectrum of features, the first three together apparently being analogous to delta moraines in lacustrine settings.

7.7.1 Morainal banks

Morainal banks form by a combination of lodgement, melt-out, dumping, push and squeeze processes when a glacier terminus, grounded in water, is in a stable or quasi-stable state (Powell 1983, Powell & Molnia 1989). (Note that the American term morainal in this context is well

Bore-hole log   Type of flow deposit

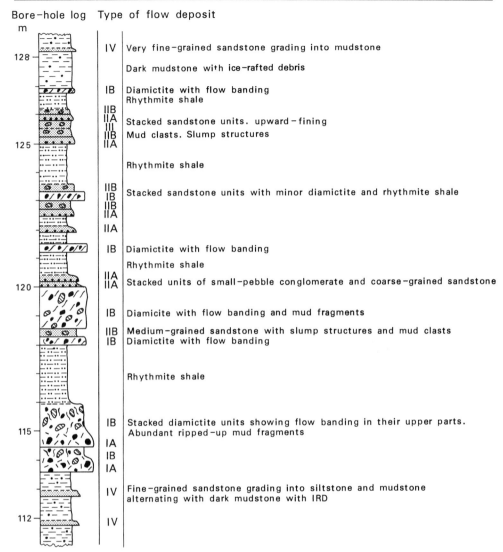

Figure 7.9   Part of the Late Palaeozoic Dwyka Formation, Kahari Basin, southern Africa, recovered in a borehole, illustrating the relationship between different types of flow deposits (adapted from Visser 1983b, with permission of the International Association of Sedimenologists).

established in the literature, so is used in preference to the British adjectival form morainic.) These banks are often intimately associated with submarine outwash. Depositional processes are further complicated by the partial lifting of the ice margin from the bed as the tide comes in, and by tidal-pumping, allowing subglacial meltwater to influence growth of the bank. Morainal banks are composed of pockets of diamicton enclosed in poorly sorted sandy gravel (with clasts up to boulder size) or gravelly sand. The foreslope of the bank is affected by resedimentation processes such as sliding, slumping and gravity flowage. The up-glacier side of the bank may also collapse once the glacier has receded.

In McBride Inlet, a short arm of Glacier Bay in Alaska, Powell (1983) observed that a morainal bank formed between 1978 and 1980 (Fig. 7.10). Within this period, 1.2 million cubic metres of sediment were deposited at a rate of $400000\,\mathrm{m^3yr^{-1}}$. Since the early 1980s the top surface has been partly exposed at low tide. The flanks of the bank slope at $15°$ (foreslope) and $9°$ (backslope).

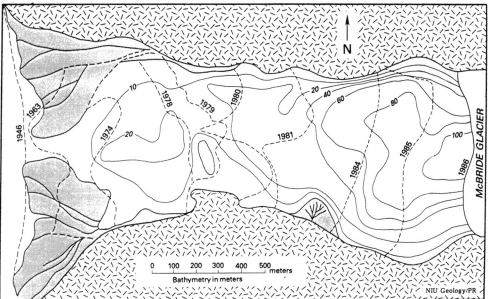

Figure 7.10 McBride Inlet, Glacier Bay, Alaska, showing the morainal bank at the entrance (now exposed) and a younger morainal bank (between the 1978 and 1980 glacier fronts) that is capped by submarine outwash sediments. A small delta on the south side occurs near the 1984 position (map from Powell & Molnia 1989, with permission of Elsevier Science Publishers, Amsterdam).

7.7.2 Grounding-line fans

Also known as subaqueous (but more properly subaquatic) or submarine outwash fans, **grounding-line fans** extend from a glacier tunnel that discharges subglacial meltwater where a glacier terminates in the sea (Powell 1990; Fig. 7.11). The bulk of the sediment is released very rapidly from meltwater, which issues as a horizontal jet at, or near, the sea floor during full-pipe-flow (Fig. 7.4). This outwash contains a range of particles from coarse gravel to mud, the coarser fraction being deposited almost immediately. The rest of the fan system includes sediment gravity-flows, slumps and deposits from the plume of turbid meltwater as it rises above the bed. Much of the fine material is carried away to more distal settings in the sediment plume.

Figure 7.11 Grounding-line fan developing in front of Harriman Glacier, Prince William Sound, Alaska. This type of feature, developed close to or just below water level, represents a platform that facilitates the advance of the glacier.

The complexity and nature of grounding-line fans, which are difficult to investigate in modern settings, is demonstrated by parts of the Late Palaeozoic Dwyka Formation in South Africa (Visser et al. 1987; Fig. 7.12) and the Late Palaeozoic Wynard Formation in Tasmania (Powell 1990). In the latter, six main facies were recorded.

Figure 7.12 Grounding-line fan complex in the Late Palaeozoic Dwyka Formation, Kransgat River, Karoo Basin, South Africa. (a) Massive diamictite at the base, overlain by interbedded sandstones and diamictites that show signs of slumping. (b) Close-up view of slumped sandstone and diamictite. (c) Soft-sediment sandstone dyke penetrating diamictite, the result of rapid burial and fluidization of an overlying sandstone layer. (d) Stratified diamictite with large ice-rafted boulder, representing more distal facies compared with (a).

(a) Stratified conglomerates with rounded to subrounded clasts and trough cross-bedding. This facies was interpreted as representing a recession of the glacier.

(b) Clast-rich stratified diamictite, with subangular to subrounded clasts, many of which are striated, and with striated cobble pavements, representing a quasi-stable terminus.

(c) Deformed pebble conglomerate and coarse sandstone, with flow structures resulting from loading on to the diamictite and ice-push as the fan prograded, followed by flowage as the ice support was lost when recession took place.

(d) Trough cross-bedded sandstone and conglomerates, with deformation and injection of diamictite resulting from ice-push during continued progradation of the fan.

(e) Cross-bedded sandstone and poorly sorted pebble/boulder conglomerates, with a few striated clasts, also formed during progradation.

(f) Stratified diamictite with 5–50% subangular to subrounded clasts, including sandstone rafts incorporated during soft-sediment deformation.

Geometrically, the better stratified parts of the fan have a barchanoid (concave down-flow) form.

Grounding-line fans that develop during sustained advances are not well preserved because of reworking. Nor are fans that form during rapid recession well preserved because there is insufficient time for them to build up. Like morainal banks, they are best developed if the ice margin is quasi-stable.

7.7.3 Ice-contact deltas

Developing out of grounding-line fans, when a glacier becomes more-or-less stable for tens of years or more, are tidally influenced fan deltas (Powell & Molnia 1989, Powell 1990), alternatively known as proglacial fan deltas (Fig. 7.8). The fan grows by direct addition of sediment from submarine outwash-streams and from sediment gravity-flows at the delta front. Mass-flows elsewhere along the glacier front may contribute to the growth of the fan. As the glacier pushes against the delta, the sediment is subjected to folding and thrusting, processes which tend to increase the height of the delta (Fig. 7.4). Such features are known as **ice-contact deltas**. With continued re-advance, the glacier rides up the back of the fan and deposits lodgement till, and perhaps also forms typical morainal bank accumulations. If the terminus then recedes, basal ice may become buried. Tunnels, through which streams discharge, become inclined up-glacier and prone to blockage, promoting occasional outbursts of sediment-charged water. During recession much of the coarser load is dumped on the reverse side of the fan, and backsliding and collapse of sediment is likely. The larger a fan grows, the more stable it becomes, providing a platform over which subsequent advance can take place (Fig. 7.11).

In Alaska, ice-contact deltas form extremely rapidly. A good example is that formed during the recession of Riggs Glacier in Glacier Bay (Fig. 7.8). In 1979, the glacier terminated in about 55 m of water, but by 1985 the delta plain had aggraded above high-tide mark and extended 100 m further into what had been 32 m of water in 1981; a total of one million cubic metres of sediment had thus accumulated in only four years.

7.7.4 Fluviodeltaic complexes

Where a river or stream enters the fjord directly, especially if carrying a large bedload from a glacier up stream, a delta with a braided top may develop (Fig. 7.13). During peak melting in early summer the bedload comprises material up to cobble size, which is deposited both as a topset and a foreset, together with finer gravel and sand. Channel filling and switching gives rise to trough cross-bedding often on a scale of tens of metres wide and several metres deep. Active channels may extend down the delta slope and on to the fjord floor; levées and terraces may also develop along the sides of the channels. One such channel in Glacier Bay (Queen Inlet) is 32 m deep and up to 259 m wide, extending for 10.5 km (Syvitski 1989). Slumping is

Figure 7.13 Fluviodeltaic complex in Kejser Franz Josef Fjord, East Greenland (foreground). Note the plume of sediment, derived from a glacially influenced river flowing from the right.

also a common process on the delta slope. Finer material will mostly be carried away in suspension, but some may be deposited between distributary channels in intertidal areas. The top of the delta front may be reworked by waves and tides, creating beach ridges. An example of a fluviodeltaic complex in Svalbard is illustrated in Figure 7.14a. The principal facies recorded, together with their interpretations are, from top to bottom, as follows:

- *sandy gravel* – forms topset of stratified pebbles and cobbles, mostly of subrounded and subangular shapes (Fig. 7.14b); fluvially deposited on the delta top;
- *sand, gravelly sand and sandy gravel* - cross-bedded on a metre scale; fluvially derived, subaquatically deposited delta foresets (Fig. 7.14c);

Figure 7.14 Dissected, elevated Weichselian fluviodeltaic complex at Engelskbukta, Spitsbergen, showing characteristic facies: (a) general view of delta, showing low-angle foresets dipping seawards (to the right); (b) poorly sorted sandy gravel, forming a topset about 1 m thick; (c) outermost foresets in delta, comprising interbedded sand and gravel; (d) gently dipping, well sorted, stratified sand with ice-rafted clast.

212

- *sand* – graded coarse to fine sand or silt beds, a few centimetres to decimetres thick, with dispersed gravel lonestones (Fig. 7.14d); turbidity-current-derived sediment with ice-rafted dropstones;
- *mud* – massive blue clayey silt and clay, with whole or broken mollusc shells and dispersed stones; fjord-bottom mud of suspension origin with minor ice-rafted debris;
- *diamicton* – massive gravelly sandy mud, with shell fragments with up to 30% gravel clasts; waterlain or flow till deposited on the fjord floor.

7.7.5 Proglacial laminites

Laminites have been studied most thoroughly in Alaskan fjords, where they lie beyond the morainal banks and ice-contact deltas, but still in a proximal position in relation to the glacier (within 1–2 km) (Powell & Molnia 1989). Laminites are produced by deposition of suspended sediment from streams, tidal and wind-generated currents, subaquatic slumps and gravity-flows. Variations in stream discharge, combined with the effect of tidal currents, frequently produce rhythmically laminated sediments called **cyclopsams** (graded sand/mud couplets) and **cyclopels** (silt/mud couplets) (Mackiewicz et al. 1984).

Experiments using traps in McBride Inlet, Glacier Bay (Cowan & Powell 1990) have shown that cyclopsams and cyclopels are derived from turbid overflow plumes, but that deposition as laminae is related to tides, and two graded couplets are produced each day. The sand flux is greatest during low tide, and lowest near high tide. Small-scale gravity-flows may interrupt

Figure 7.15 Fjord-bottom sand/silt rhythmites (probably tidally related), associated with tidewater Comfortlessbreen, Engelskbukta, Spitsbergen. These sediments have been lifted above sea level by large-scale glaciotectonic processes.

the laminite sequence. Other laminae may be produced as a result of variations in discharge on a diurnal basis. Pleistocene cyclopsams and cyclopels have been recognized in other fjord areas with lower tidal ranges than Glacier Bay, ranging from temperate latitudes (e.g. Whidbey Island, Washington State; Domack 1984) to high Arctic latitudes, such as Ellesmere Island and Svalbard (Fig. 7.15), although these may be daily or seasonal couplets, rather than tidal ones.

Cyclopsams are interbedded or interlaminated with coarser outwash sediments in the ice-proximal zone. Cyclopels also form in a proximal position during periods of low discharge or in quiet areas adjacent to the glacier front. Beyond 1–2 km from the terminus, laminites are lost because all the silt has already settled out and sedimentation rates are much reduced, allowing bioturbation and flocculation to take place.

In the older geological record, tidal rhythmites have been reported, for example, from the Early Proterozoic Gowganda Formation in Ontario, Canada (Mustard & Donaldson 1987), and the Neoproterozoic Elatina Formation, South Australia (Williams 1989). Many other ancient laminites in glacial sequences are probably of this type.

Figure 7.16 Neoproterozoic fjord-bottom rhythmites with abundant ice-rafted material, Henan Province, China.

An important additional component to laminites is often represented by **dropstones** (or **lonestones** if a non-genetic term is preferred), providing one of the most reliable criteria for recognizing glacial influence in the geological record (Fig. 7.16). Dropstones of ice-welded diamicton or mud may also be present; these are known as till pellets, although some of them may reach boulder size. Dispersed clasts frequently exceed the thickness of the laminae, and show disruption of the laminae beneath, with draping of laminae over the top.

7.7.6 Fjord-bottom sediment complexes

Further from the glacier, and interstratified with the proglacial laminites, is homogeneous mud derived from glacial rock flour. This mud contains variable amounts of iceberg-rafted debris and therefore may grade into diamicton (more than about 1% gravel). Such a deposit has been referred to as **bergstone mud** (Powell & Molnia 1989), and its extent depends on the paths taken by icebergs. Dropstone structures are not visible in these homogeneous sediments.

7.7.7 Beach and tidal-flat features

Beaches are best developed close to areas where there is, or has been, an abundant supply of subglacial outwash sediments. Often beaches consist of coarse gravel, becoming sandy down-drift of the outwash streams. Beaches also form in sheltered rocky embayments. Spits commonly develop across river outlets, and these may be backed by lagoons and marshes (Fig. 7.17).

Tidal flats are generally of fine sand and mud, often dotted with iceberg-rafted debris. Tidal flats are common in areas of quiet water, towards the margins of river outlets; they often form on top of deltas, after the main distributaries have moved elsewhere.

Figure 7.17 Spits and lagoons, containing reworked glacial and glaciofluvial sand and mud, Engelskbukta, Spitsbergen. Comfortlessbreen is in the background (cf. frontispiece).

Beaches and tidal flats in glacier-influenced fjords show a range of features arising from iceberg action as the tides rise and fall, although in the polar regions features produced by icebergs may be difficult to distinguish from those resulting from sea ice. Icebergs rapidly break down into **bergy bits**, ice blocks several metres across. Tidal movements may drag an iceberg across the beach or tidal flat, creating a groove with levées (Fig. 7.18a). Grooves run up and down the beach, or across it if longshore drift takes place. Often grooves are irregular in direction and cut across one another. Icebergs may also be rocked by waves, creating wallows, and others may push up ridges as the tide comes in (Fig. 7.18b). If icebergs are not well grounded, then various bounce-, chatter- and rollmarks may form. Some icebergs become stranded in intertidal zones, and on sandy or muddy shores may become buried, creating so-called iceberg rosettes or a pitted surface when they melt (Fig. 7.18c). Dirty icebergs, comprising basal ice, may also become buried, leaving behind diffuse pods of till or mud in the beach deposit.

If strong tidal currents and waves winnow out fine material from the surface of boulder- or cobble-rich diamictons in beaches, icebergs may compress the boulders into the matrix, creating a flat surface or type of gravel pavement (Fig. 7.18d). Fine, irregular striations may form as the ice moves over the pavement.

7.7.8 Iceberg-turbate deposits

Icebergs may become stranded on shoals, disturbing the sediment and creating an **iceberg turbate**, although there seem to be few descriptions of such features in the literature. Since shoal deposits are typically winnowed diamicts, the turbate will have gravel as the dominant component. If it is interbedded with other facies, iceberg grounding may be evident in the form of deformation structures. If the icebergs contain debris, the resulting sediment will be of mixed composition and have a disorganized fabric, often building up into irregular ridges.

7.8 Stratigraphic architecture in fjords

To the author's knowledge, no systematic studies, combining deep drilling through fjord sediment to bedrock and seismic surveys, to determine the stratigraphic architecture of a modern fjord have been undertaken, although it has been possible to gain an appreciation of the structure of bottom sediments in some Arctic fjords using geophysical techniques combined with shallow coring.

In Arctic fjords, where sedimentation rates are low compared with Alaska, high resolution acoustic profiling has, in some places, been able to penetrate to bedrock. As an example, the work of Sexton et al. (1992) in Krossfjorden, Spitsbergen may be summarized. Krossfjorden is a fjord currently influenced mainly by non-surge-type, grounded tidewater glaciers. Along with its branches, it has several distinct basins (Fig. 7.19a). Seismic surveys have yielded details of the stratigraphic architecture of the sediment that has accumulated since

216

Figure 7.18 Features formed on beaches in iceberg-influenced fjords. (a) Grooves on sandy gravel beach created by icebergs from Columbia Glacier, Prince William Sound, Alaska. (b) Push ridges on beach formed by icebergs at high tide, Columbia Glacier, Alaska. (c) Iceberg rosettes, pits resulting from burial of sea or glacier ice by beach material, followed by slow melting of the ice, Engelskbukta, Spitsbergen. (d) A gravel pavement formed by ice-compaction and tidal-current winnowing of shoreline sediments, Prince William Sound, Alaska.

(a)

(b)

(c)

(d)

(a)

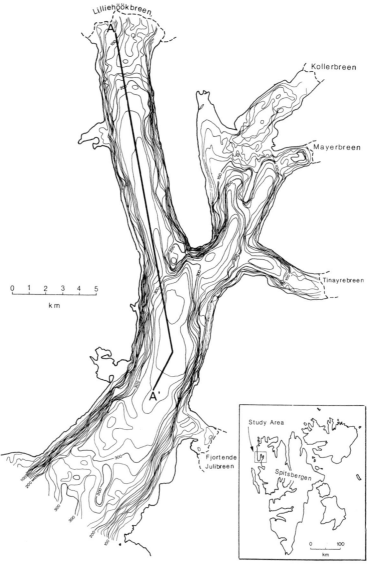

(b)

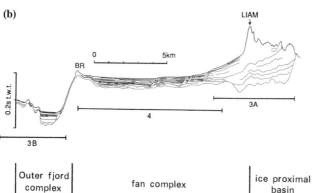

Figure 7.19 Krossfjorden, NE Svalbard. (a) Bathymetric map (20 m interval contours), showing multiple basins and termini of tidewater glaciers influencing sedimentation. (b) Acoustic profile (3.5 kHz record) and interpretation illustrating seismic stratigraphic architecture in Lillie-höökfjorden, the northwestern arm of Krossfjorden. LIAM is the Little Ice Age moraine and BR represents a bedrock ridge and break of slope at the mouth of this arm of Krossfjorden (from Sexton et al. 1992, with permission of Elsevier Science Publishers, Amsterdam).

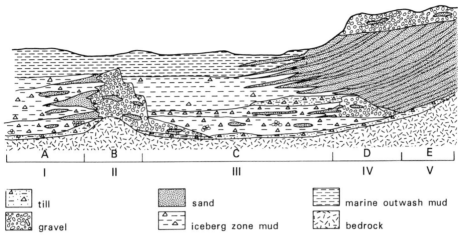

Figure 7.20 Hypothetical facies architecture in a glaciomarine–fjord complex based on investigations of Alaskan tidewater glaciers (simplified from Powell 1981).

the fjord system was fashioned by the ice. Up to 180 m of unlithified sediment above bedrock is present. The acoustic character of the sedimentary pile is variable, and three main complexes have been distinguished (Fig. 7.19b).

- *Ice-proximal unit* – interpreted as subaquatic, hummocky terminal moraines, formed annually during a general recession, inside the Little Ice Age maximum; the material is a fine-grained diamicton with no internal structure.
- *Fan complex* – comprising structureless till or a sheet of glaciomarine sediment at the base, overlain by well laminated sand and silt; this glaciomarine fan comprises material that rained out from suspension or was deposited from debris-flows or turbidity currents; the complex formed as the late Weichselian glaciers receded up-fjord.
- *Outer fjord complex* – comprising a blanket of both structureless and laminated material (depending on the degree of bioturbation), and a hummocky till above bedrock, similar to the ice-proximal unit mentioned above; this complex formed during and following the late Weichselian glacial maximum.

Surveys of some Baffin Island fjords confirm the broad pattern found in Spitsbergen fjord basins of a glaciomarine mud-draped till, but, in addition, illustrate the importance of resedimentation and Coriolis effects in piling up masses of sediment to one side of the fjord (Syvitski 1989, Boulton 1990). In Alaskan fjords, the thickness and internal structure of sedimentary sequences have also been determined using high-resolution profiling methods, although the sedimentary facies that make up the several hundred metre thick sequences are not known directly. However, Powell (1981) has developed a model of a hypothetical sedimentary sequence (Fig. 7.20) that may arise from the progressive development of the five facies associations he defined (§ 7.6.1), and which illustrates at least a little of the complexity of the stratigraphic architecture that may develop in a fjord.

8 Glaciomarine processes and sediments

Part 2 Continental shelf and deep sea environments

8.1 Introduction

High-latitude continental shelf areas under the influence of glacigenic processes are particularly sensitive to changes in climate. Ice sheets that border these continental shelves, not only control sea level, but also generate cold bottom-water that influences the oceans throughout the world. The problems and cost of access to the continental shelf glaciomarine environment – represented first and foremost by Antarctic ice shelves, outlet glaciers, ice tongues and ice cliffs – have so far limited the acquisition of knowledge concerning these areas. Nevertheless, a considerable body of data has been obtained during the past decade, and reviews have stressed the importance today of glaciomarine processes on high-latitude continental shelves (e.g. Andrews & Matsch 1983, Dowdeswell 1987, Drewry 1986, Hambrey et al. 1992).

Ten per cent or, possibly, much more of the world's oceans and continental shelves receive glaciomarine sediment today (Drewry 1986). Around Antarctica, icebergs influence sedimentation, albeit to a volumetrically minor degree, up to a distance of several hundred kilometres from the continent (Fig. 8.1a). The northern iceberg limit extends to the southern tips of South America, South Africa and New Zealand, and there were reports during the 19th cen-

Figure 8.1 Decaying icebergs off Dronning Maud Land, East Antarctica. (a) Tilted, tabular iceberg grounded on the outer continental shelf. (b) Iceberg in advanced stages of decay close to the coast. Note the prominent debris band, probably originally incorporated at the base of the ice mass, and the ice shelf in the background.

tury of icebergs creating a hazard to shipping around Cape Horn and the Cape of Good Hope. Because of their favoured preservation potential, continental shelf sediments are the most common of pre-Pleistocene glacigenic sequences, and many detailed accounts have been published in recent years.

One of the earliest and most stimulating attempts at reconstructing the glaciomarine environment, typified by Antarctica and the margins of the Pleistocene ice sheets of the Northern Hemisphere, was that of Carey & Ahmed (1961). The model developed by these authors was widely used, but was founded on few field data and it seriously overstressed the rôle of ice shelves as providers of sediment to the marine environment. Recent reconstructions are founded on a much better database, although this is still inadequate in view of the wide range of glaciological conditions that occur at the margins of marine ice masses. Relatively poorly known are the modes of debris transport by ice to the continental shelf, the thermal regime of the transporting glaciers, the rates of sediment supply, the importance of meltwater, the rôle of water depth and ocean currents, the character of sediments deposited beneath floating ice and their modification by marine processes, and the rôle of sea ice. Nevertheless, we now have a good idea of the range of processes that influence sedimentation in these settings, and they form the basis of this chapter.

8.2 Transport of glacigenic sediment out of the ice-sheet system

8.2.1 Nature of sediments arriving at the grounding zone

The volume and distribution of sediment within ice masses are largely controlled by ice dynamics and thermal regime (see Drewry & Cooper 1981, Drewry 1986, Dowdeswell 1987, for more detailed reviews). As noted in Chapter 7, the thickness of the basal debris-rich zone varies according to glacier type. This debris zone is derived as a result of erosion at the base of a sliding ice mass and entrainment by regelation of dirty ice. Ice masses which are predominantly cold tend to have thicker basal debris layers than do temperate glaciers; commonly this layer is 10 m or more thick where terminating on land, but it is not known for certain whether similar thicknesses prevail where ice masses enter the sea. From coring and radio-echo sounding of ice sheets, the thickness of the basal layer tends to be of the order of 1% of the total thickness, so some Antarctic glaciers may have as much as 100 m of basal debris when they reach the grounding-line. Sediment-rich ice occurs in discrete bands (Fig. 8.1b), but the concentration is variable, ranging from 0.01 to 70%, or about 1% on average, showing an irregular but approximately exponential decrease upwards. This distribution and density pattern is important from the point of view of determining the rate of sedimentation once the ice mass begins to float.

Other sediment sources, volumetrically, are relatively minor around most ice margins that border continental shelf areas. Exceptions are the few instances in Alaska where glaciers terminate in the open sea; here supraglacial debris is also important.

8.2.2 Ice–ocean sediment transfer routes

Excluding lodgement, melt-out and shearing of the debris layer beneath the grounded ice sheet, debris is transferred out of the Antarctic ice-sheet system in three ways (Drewry & Cooper 1981).

Via ice shelves
On entering deep water, grounded ice on a broad front decouples from the bed and floats out over the sea, where it undergoes creep-thinning as a result of acceleration following the reduction of basal and lateral drag (Fig. 8.2). For sedimentation, it is important to consider mass balance and particle paths. Most ice shelves receive considerable snow accumulation, so particle paths are inclined downwards as the accumulation increases seawards. If there is net basal melt, particle paths intersect the bottom of the ice shelf, and any debris in transit in the basal layer is likely to be melted out prior to calving at the ice front. So sedimentation today is important at the inner margins of a continental shelf near the grounding-line, but of little significance on the outer shelf and in the open ocean. However, freeze-on of oceanic ice beneath some ice shelves is important. Beginning close to the grounding-line, freeze-on may prevent all of the basal and englacial debris from being released. The remaining debris is therefore transferred to the ice-shelf edge and deposited in the open sea from icebergs.

Figure 8.2 Embayment in the Brunt Ice Shelf, off Halley Station, East Antarctica. RRS *Bransfield* is anchored against fast ice.

Via ice cliffs
Where a grounded Antarctic ice margin calves directly into the sea, ice cliffs form (Fig. 8.3). Wave action at the foot of the cliffs and the slow rate of advance of the cliffs promote sedi-

Figure 8.3 Grounded ice cliff near Barnes Glacier, Ross Island, Antarctica, with winter fast ice at its foot.

mentation close to the ice edge prior to calving. In the Arctic, extensive ice cliffs are found only on Nordaustlandet in the Svalbard group of islands.

Outlet glaciers and ice streams
Inland ice draining towards the coast is channelled into ice streams or outlet glaciers, the latter being constrained by rock walls. Such glaciers generally have a high rate of discharge and may extend out into the sea as ice tongues. Observations of calving icebergs from such glaciers show that they retain a basal debris layer up to 15 m thick, with variable amounts of supraglacial debris. The sediment-release process is similar to that under ice shelves, but, in addition, they produce an almost continuous supply of small, debris-rich icebergs. The bulk of the glacigenic sediment in the Southern Ocean today probably comes from outlet glaciers.

The relative importance today of these three modes of transfer of ice from Antarctica have been estimated to be as follows (Drewry & Cooper 1981):

Ice shelves	62%
Ice cliffs	16%
Outlet glaciers and ice streams	22%

However, from a sedimentological point of view, there are significant differences in the locations and volumes of sediment produced by these different glacier types (Table 8.1).

Ice shelves and outlet glaciers respond rapidly to oceanographic and climatic changes. Their grounding-lines thus migrate frequently across the continental shelf, producing thick complexes of strongly diachronous sediment (Drewry & Cooper 1981).

Table 8.1 Variations in the location and relative proportions of different types of entrained sediment, according to type of ice mass (after Drewry 1986).

Glaciological regime	Debris type		
	Basal	Englacial	Supraglacial
Outlet glaciers			
via mountains	• • • •	• •	• •
ice stream	• • • •	•	•
Ice cliffs			
near mountains	• • •	• •	• •
ice sheet edge	• •	•	•
Ice shelf			
from ice sheet	• •	•	•
via mountains	• • •	• •	•

Sediment quantity			
trace	small	moderate	substantial
•	• •	• • •	• • • •

8.3 Sedimentation on glacier-influenced continental shelves

The range of sediment transport paths, processes of deposition and resulting sedimentary facies are summarized in Figure 8.4 for an ice shelf in recessed and advanced conditions.

8.3.1 Delivery of sediment to the glaciomarine environment

Ice dynamics

As much as a half of the Antarctic ice sheet was considered by Zotikov (1986) to be at the pressure melting point at its bed, a view supported by clear evidence from radio-echosounding of subglacial lakes beneath the thicker parts of the ice sheet (Oswald & Robin 1973). Under these conditions, basal sliding is bound to generate a debris layer. Although this basal debris layer is normally visible only in a few places around the margin of the continent (Drewry 1986), it is believed to be widespread in the interior of Antarctica. Deposition of the bulk of this sediment in the marine environment is by subglacial melting on the continental shelf beneath an extended grounded ice sheet or close to its grounding-line. Melting of icebergs provides a more limited source of debris in the open ocean.

Another mechanism of transport of basal debris to the marine environment has been identified beneath ice streams hundreds of metres thick draining the West Antarctic ice sheet and feeding the Ross Ice Shelf (Shabtaie & Bentley 1987, Alley et al. 1989). Both theory and seismic velocity data support the view that the fast-flowing ice streams are moving over a water-saturated bed of till (or perhaps previously deposited glaciomarine sediment) and deforming it, thereby transporting it seawards. Transport of this material ends as the ice stream merges with the ice shelf and achieves buoyancy, but at the same location melting of the underside of the ice stream by currents beneath the ice shelf releases additional basal debris. These processes of debris transport and release are followed by gravity-flow and slump pro-

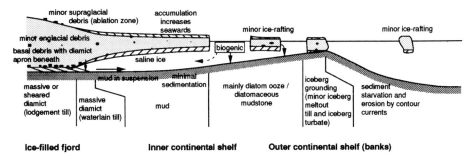

(a) Ice shelf in recessed state

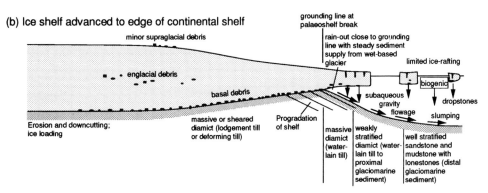

(b) Ice shelf advanced to edge of continental shelf

Figure 8.4 Ice dynamics, sediment sources and sedimentary processes, products and their interpretation at the margin of the Antarctica ice sheet. (a) Ice shelf with grounding-line at the inner part of the continental shelf, as at the present day. (b) Ice shelf having grounded on the continental shelf and advanced across it to the continental shelf break, where it becomes decoupled from the bed (glacial maxima) (from Hambrey et al. 1992, with permission of Gebrüder Borntraeger, Stuttgart).

cesses to create, beneath the inner margin of the ice shelf, a prograding complex of glacially derived sediment. Alley et al. (1989) described the resulting feature as a **till delta**, although the non-genetic term **diamict apron** is preferred here. This type of prograding wedge, but on a much larger scale, has been recognized in seismic records from the Antarctic shelf for some time (Hinz & Block 1983), and is now considered by some to be characteristic of ice-sheet deposition (Cooper et al. 1991a). It is conceivable that this process of diamict apron formation was the primary mode of continental shelf progradation when glaciers reached the shelf break, as well as being, on a smaller scale, the main mechanism for deposition on the inner shelves today (Hambrey et al. 1992).

Debris can also be transported into the marine realm after falling or being blown on to the glacier surface. Rock debris and wind-blown sand are widely observed on the surfaces of outlet glaciers passing through the Transantarctic Mountains in southern Victoria Land (Barrett et al. 1983), and are probably equally common in other localities where ice passes through coastal ranges or where nunataks are exposed. In only a few places, however, is supraglacial debris sufficiently abundant to be evident in aerial photographs or satellite images, and it is probably unimportant as a proportion of the total sediment, except where the sediment supply from other sources is exceptionally low. However, aeolian sediment from these sources can be spread as far as the northern limit of icebergs because it is carried on top of, and within, the ice. Such debris can be readily recognized from its well sorted fine-grained nature (aeolian sand) and angular shape (scree).

In the Arctic, rockfall debris and aeolian sand on icebergs derived from glaciers and ice caps terminating at the open coast (e.g. northeastern Svalbard) are of minor importance. Only in Alaska, where a few glaciers approach the open coast, is supraglacial transport locally important. However, during Pleistocene glacial periods, when ice descended from the lofty mountains to the open coast on a broad front, the transfer of supraglacial sediment was highly significant, not only in Alaska, as demonstrated by very thick Late Cenozoic sequences that have been tectonically uplifted, but also off northwest Europe, British Columbia and the southern Andes.

In summary, the main source of glacial debris for the open marine glacial regime in Antarctica is attributed to outlet glaciers and ice streams (Drewry & Cooper 1981, Orheim & Elverhøi 1981). However, most of this debris is probably deposited close to the grounding-line, from which position it may be further distributed (and modified) by gravity-flows or currents. Debris deposited from icebergs is volumetrically minor compared with that originating from other sources. This, too, has the attributes of basal glacial debris. Such debris may occur far from the source if the iceberg turns over before the basal debris has melted out. Ice-rafted debris in the stratigraphic record can provide important evidence of the existence of floating ice at sea level.

Glacial meltwater input

Sediment in suspension and bedload can enter the marine environment directly from supraglacial and subglacial meltwater channels in tidewater glaciers, floating ice tongues and ice shelves, or from terrestrial streams. In the Antarctic the latter are small and rare, with maximum discharges of a few cubic metres per second. According to Kotikov (1986), subglacial meltwater is generated at a rate of several cubic kilometres per year beneath the Antarctic ice sheet and its outlet glaciers. However, this volume is insufficient, or the water inadequately channelized, to generate visible turbid sediment plumes in surface waters around Antarctica. Some subglacial transport of fine-grained sediment is suspected from the abundance of terrigenous sediment in deep-water mud accumulating around the continent today. In contrast, meltwater and associated suspended sediments have been observed immediately seawards of some glaciers at the northern tip of the Antarctic Peninsula, where temperatures are significantly higher (Griffith & Anderson 1989).

Glacial meltwater is much more significant in the Northern Hemisphere, where cold ice fronts are grounded on the continental shelf. For example, the ice caps on Nordaustlandet in Svalbard, which together provide the longest continuous sea frontage of any Northern Hemisphere ice mass (120 km), reveal discrete subaquatic stream outlets, as indicated by sediment plumes.

Sea ice and river ice

Terrigenous debris entrained into sea ice occurs in littoral environments by freeze-on, as described in Chapter 7. This process is facilitated if extensive shallow areas exist, as is the case in the Arctic, but in the Antarctic such conditions are relatively rare. Ice-rafting of littoral debris that becomes anchored to the bottom of sea ice, and of fine sediment that may be incorporated in sea ice when turbid water freezes, is of major importance in the Arctic Ocean and surrounding seas, in areas far removed from any glaciers. Indeed, muds derived from sea ice are thought to form about 80% of the Late Cenozoic sediments in the central Arctic Ocean (Clark & Hanson 1983).

Sea ice also plays a rôle in the sedimentation of diatom ooze in Antarctica. Diatom blooms occur beneath the sea ice as it breaks up in summer, but the diatoms become trapped in the ice when new platelet ice forms beneath the floes. Often the ice floes are strongly coloured, giving rise to so-called **brown ice**.

Arctic rivers discharge large volumes of ice into the ocean during the spring break-up, allowing river gravels and sands to be ice-rafted off shore. To distinguish these types of ice-rafting sediment from iceberg sediment is of major importance in assessing the palaeoclimatic significance of dropstones in sediments, for example, around the Arctic Basin.

Sea and river ice-rafted debris has been reported from some pre-Pleistocene sequences, even for times when there is no other evidence of glaciation, e.g. the Jurassic and Cretaceous of the Canadian Arctic, Spitsbergen, Siberia and Australia (Frakes & Francis 1988).

The transport of aeolian sediment on to sea ice and its subsequent release into the sea is an important process in some polar areas. For example, Barrett (1986) recorded deposition of up to $1\,\mathrm{mm\,yr^{-1}}$ of aeolian sediment in McMurdo Sound, offshore from the Dry Valleys in Antarctica (Barrett et al. 1983)..

Biogenic material

The polar seas are biologically rich in those areas where the sea ice breaks up in summer. Upwelling of nutrient-rich bottom-waters and low clastic sediment input, facilitate the production of a diverse fauna and flora. Siliceous ooze is the principal component around Antarctica, where a zone several hundred kilometres wide, made up largely of diatom fragments with a minor ice-rafted terrigenous component, rings the continent. To the north lies calcareous ooze comprising coccoliths and foraminifera, while the transitional zone coincides with the boundary between cold Antarctic waters and the warmer oceans to the north (the zone being referred to as the Antarctic Convergence). The importance of diatomaceous sediment has been also been recognized near the Antarctic coast. For example, Domack (1988) found a predominance of diatom ooze just outside the fjords of the Antarctic Peninsula, as did Dunbar et al. (1989) in McMurdo Sound. The dilution of the ooze by terrigenous sediment depends on the proximity of outlet glaciers. On the other hand, areas such as the inner Weddell Sea, which have a more permanent and relatively dense sea-ice cover, have little diatomaceous sediment. Some diamictites dating back to Oligocene time, recovered from the deep drill holes, also contain significant amounts of diatoms (Barrett 1989a,b, Barron et al. 1989). Nearly pure oozes with only minor ice-rafted material are present in many places near the Antarctic coast. The distribution of siliceous biogenic material on the inner Antarctic shelf is strongly influenced by marine currents, and transport is generally in a westerly direction.

The distribution of diatomaceous sediments in the Arctic is less clearly defined, and to a large extent reflects the permanancy of the sea ice. For example, the relatively ice-free coasts of western Svalbard have abundant diatomaceous sediment, whereas the Arctic Ocean, with its near-complete sea-ice cover, does not.

Other biogenic components (e.g. foraminifera, calcareous nanofossils, radiolaria, bryozoa, siliceous sponges, invertebrate fossils) form only a fraction of a per cent of the total sediment, but their abundance, along with diatoms, demonstrates the high productivity of waters in polar regions.

8.3.2 Ice–ocean interactions

The nature of the interaction between glaciers and the sea on continental margins is relatively poorly known. In particular, the processes operating just seawards of the grounding-line, where tidal pumping may be important, and beneath floating ice masses, are poorly known, and glacial sedimentologists await the development of new technology that will allow investigation of these inaccessible areas by remotely operated vehicles or drilling through the ice. The most important ice–ocean interactions have been outlined by Drewry (1986) and Dowdeswell (1987), and are summarized below.

Rates of ice-cliff melting
The rate of debris production from grounded ice cliffs is dependent on the ice velocity, the temperature difference between the water mass and the ice front, the velocity of any current, and the length of contact between the ice front and the water mass. For ice margins terminating in shallow water, surface warming of the sea in summer accounts for the loss of significant amounts of ice.

Melting and freezing at the base of ice shelves
The thermal regime at the base of an ice shelf is important to glaciomarine sedimentation because it controls whether all or most of the basal debris is released at the grounding-line through melting, or whether some of it is transferred to the open ocean by being protected by the freezing-on of a layer of oceanic ice. Ice floating in water is prone to melting because heat is transferred from the warmer liquid into the ice. The process beneath an ice shelf is facilitated by water circulation in the form of tidal currents and meltwater plumes. The geometry of the underside of the ice also helps, since this slopes downwards towards the land; ice shelves typically thin from around 1000 m near the grounding-line to 200–500 m at the seaward limit. The most vigorous circulation is near the grounding-line, thought to be in the form of convective turbulent flow linked to "tidal breathing", so it is here that most debris melts out.

In certain circumstances, basal freeze-on occurs. If deep-flowing saline water close to the freezing point rises as the grounding-line is approached, its freezing point is raised due to lowering of pressure, and it may freeze as it makes contact with the underside of the ice shelf.

Towards the outer part of the ice shelf, however, currents advect warmer water from beyond the ice shelf and undermelt occurs. Therefore, the basal freezing process is most active within a short distance of the grounding-line.

From observations on the Ross and Amery ice shelves, it is known that sedimentation is highly sensitive to the rate of melting. Thus, ice shelves represent barriers to the movement of basal glacial sediment from the continent offshore. Basal freeze-on is known to take place beneath ice shelves, including the Ross Ice Shelf (Zotikov 1980), the Amery Ice Shelf (Budd et al. 1982) and the Filchner–Ronne Ice Shelf (Thyssen 1988). These freeze-on zones may be over a 100 m thick, and represent as much as a third of the total ice shelf thickness (Fig. 8.4a). However, ice cores have demonstrated an absence of rock debris at the boundary between glacier and oceanic ice in cores through both the Amery Ice Shelf (Site G-1, Morgan 1972) and the Ross Ice Shelf (Site J-9, Zotikov et al. 1980), supporting the view that the bulk of the basal debris has indeed been lost near the grounding-line.

Iceberg calving

Changes in oceanographic conditions are often the principal factors affecting long-term iceberg production from floating glacier tongues and ice shelves. A rising sea level causes the grounding-line to lift and move inland and the ice mass to become unstable. Increased iceberg production may therefore herald a time of marked climatic warming and rapid recession of the ice shelf, a point that has to be taken into account when interpreting the climatic record from sediments with an ice-rafted component. An illustration of this is that the climatic warming in the Antarctic Peninsula this century is leading to the rapid break-up of the more maritime ice shelves and a huge increase in iceberg production. On the other hand, ice shelf and outlet glacier calving may not be related to oceanographic or climatic changes at all. Many ice shelves tend to advance at rates up to $1 \, \text{km} \, \text{yr}^{-1}$ over tens or even hundred of years before calving huge icebergs. Recent examples include tabular icebergs more than 100 km long calved from the Filchner and Ross ice shelves.

Grounded glaciers also respond to oceanographic changes, but are more closely influenced by climatic factors and mass balance. Icebergs in such cases are rarely more than tens of metres across, are irregular in shape and usually spall off the cliff face; tabular icebergs do not usually form. These glaciers also increase their iceberg production during rapid recession. Temperate glaciers that once extended across the shelf of the Gulf of Alaska often behaved in this manner during the Pleistocene Epoch.

Iceberg melting rates, fragmentation and overturning

The rate of sediment release by iceberg melting depends on the melting rate, the distribution of sediment within the icebergs, and the number of overturning and fragmentation events. Debris melting out on the surface of an iceberg may remain there until an overturning event occurs, or flow off as a slurry or be washed off by meltwater or rain, processes that cause a pulse in sedimentation. In contrast, debris on the steep sides and base of an iceberg may melt-out steadily. These processes are summarized in Figure 8.5, which documents the loss of debris from a hypothetical iceberg.

A tabular iceberg is relatively stable if its width exceeds its thickness, but melting rounds

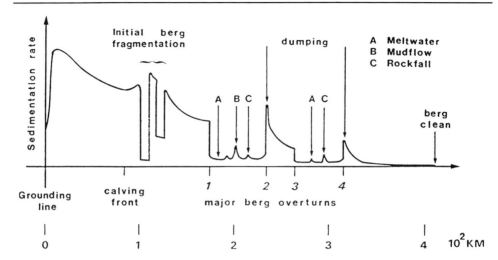

Figure 8.5 Pattern of debris release from an iceberg derived from an ice stream, and resulting sea-floor sedimentation rate. Note the various effects of calving and overturning of icebergs, the contributions from meltwater, mudflow and rockfall activity (from Drewry & Cooper 1981, with permission of the International Glaciological Society).

off the sharp edges, while waves undercut the sides at the waterline, making the cliffs prone to collapse and the production of bergy bits; as a result the iceberg becomes unstable. In an investigation of more than 2000 icebergs in the eastern Weddell Sea, Orheim & Elverhøi (1980) found that about 100–150 km offshore only 15 % had not overturned.

Rates of melting of icebergs have not been studied directly, but estimates of $10-50\,\mathrm{myr}^{-1}$ have been made for clean Antarctic icebergs by Budd et al. (1980). Dirty ice melts more slowly than clean ice, but nevertheless a dirty ice layer at the base of an Antarctic iceberg will be lost within a few years. However, any internal debris layers may survive until the iceberg finally disintegrates; normally, this is south of the Antarctic Convergence (50°–60°S), but a few icebergs drift into the warmer waters beyond.

Iceberg drift paths
The general pattern of iceberg drift is controlled by ocean currents. Thus, icebergs can move in a direction totally different from sea ice, which is more influenced by the vagaries of the wind. Around Antarctica, evidence from satellite imagery indicates that icebergs within 100–200 km of the coast tend to move in a westerly direction in the Eastwind Drift, so remaining close to the continent for up to five years. Thus, most debris is released in this coastal girdle. Those icebergs that move further offshore come under the influence of the Westwind Drift and move back eastwards. The Antarctic Convergence represents an effective northern limit for most icebergs. The average speed of Antarctic icebergs, over a year or more, ranges from 0.11 to $0.2\,\mathrm{msec}^{-1}$.

Northern Hemisphere icebergs are insignificant in size compared with their Antarctic counterparts. Most are trapped in fjords, but some distinct iceberg routes are known. The East Greenland Current flows south from the Arctic Basin and receives icebergs from major calv-

230

ing glaciers in northeastern Greenland. Most icebergs become caught up in the northbound West Greenland Current and are transported into Baffin Bay. Here they mix with icebergs that calve from the large outlet glaciers of West Greenland and the small glaciers in the Canadian Arctic. A current on the west side of Baffin Bay then carries them southwards through the Davis Strait into the North Atlantic. Some reach the Grand Banks off Newfoundland, and one sank the *Titanic* in 1912 with a loss of 1503 lives. Small icebergs from Svalbard and the Soviet Arctic islands find their way into the Arctic Basin, while larger icebergs are occasionally produced by the Ward Hunt Ice Shelf in northern Ellesmere Island. One such iceberg (ice island T3) has been tracked since 1947, and has followed an irregular, but roughly clockwise spiralling track through the Arctic Ocean.

8.3.3 Oceanographic processes

Oceanographic processes include those that are not just restricted to glaciomarine environments. However, the interaction of glacier ice and sea water influences the salinity and temperature, and thereby controls the location, rate of deposition and character of glaciomarine sediments. A fuller discussion for the Antarctic is given in Dunbar et al. (1985), but the links between oceanography and sedimentology are not well known at present.

Some processes may be considered to be particularly important in the glaciomarine environment (see Dowdeswell 1987 for further information).

Meltwater inflow and water stratification
In the Arctic or sub-Arctic, water stratification arises when meltwater from the land or glaciers enters the sea, but on the open coast, except when the winter sea-ice cover has yet to melt, currents and wind tend to mix the water. In the Antarctic there are only a few signs of meltwater flowing into the sea.

Sediment suspended in meltwater
Sediment plumes are generated where a subglacial stream discharges into the sea. Since a significant proportion of the Antarctic ice sheet is sliding on its bed, suspended sediment derived from meltwater must be inferred at grounding-lines where ice streams and outlet glaciers reach the sea and begin to float as glacier tongues or ice shelves. Sediments with abundant terrigenous mud in proximal and distal glaciomarine settings may be derived from this source. However, the process itself has not been documented.

Tides
Tides are important at grounded ice cliffs for similar reasons as in fjords (Ch. 7), although they show a less pronounced imprint on the sediment. Additionally, tides control circulation beneath floating ice masses, facilitating melting at the grounding-line and flushing out of suspended sediment.

The effects of sea ice on circulation
When sea ice forms, it leaves the underlying water preferentially enriched in brine. This brine-rich water forms a layer beneath the ice, but because it is more dense it becomes unstable. This leads to overturning and mixing of the water layers. During ice break-up, less dense fresh water is generated, inhibiting mixing and producing a relatively stable layer. Sea ice also minimizes the effect of the wind on the sea, and also inhibits mixing.

In Antarctica, brine-enriched waters accumulate on the western sides of the wider ice shelves and in depressions in the coastline. Together with surface winds, the density contrast leads to a clockwise circulation in large embayments.

Depth of water
The physical characters of the Arctic and Antarctic continental shelves differ greatly. The northern shelves are typical of most shelves in more temperate latitudes being relatively shallow and affected by a wide range of coastal processes. The Antarctic continental shelf, however, is unique in having great depth (often more than 500 m), rugged bathymetry, landward tilt and almost total glacier-ice cover at its rim. These characteristics are the result of crustal (isostatic) loading by the thousands of metres of ice on the continent. The near-absence of ice-free land and the presence of deep water immediately offshore in the higher polar areas of Antarctica mean that there is no wave-dominated coastal zone. Thus, beaches are uncommon and the rôle of sea ice in transporting sediment is much less important than in the Arctic or sub-Antarctic.

Bottom currents
Near the Antarctic continental shelf break, some of the denser shelf water flows along and down the slopes, mixing with and modifying the deep water or producing **Antarctic Bottom Water**. In some areas, such as the cold Weddell Sea, the Antarctic Bottom Water is a powerful current, winnowing, and in places deeply eroding, the sea-floor sediments, as well as having a major influence on global ocean circulation.

In the Arctic, upwelling of warmer deep waters leads to the creation of **polynyas**, areas of ocean relatively free of ice ocean in winter, as well as summer; a well known example is the Northwater, located between Ellesmere Island and North Greenland.

8.3.4 Processes of reworking

Modification of glacigenic sediments by processes characteristic of all continental margins may be a significant factor in explaining particular facies distributions. Especially important are those given below.

Current activity
Bottom currents can redistribute the finer fraction (sand and mud) of glaciomarine sediments, especially if it has been made less cohesive by bioturbation, leaving behind a lag deposit composed principally of gravel. Other bedforms, such as current ripples, or even mega-ripples

and sand waves, form at shallow depths (as on Arctic shelves) where currents are strong, and wave-induced structures also occur. Currents also transport biogenic material into zones incapable of supporting life on their own (e.g. beneath the innermost parts of ice shelves).

Subaquatic mass movement

Reworking by subaquatic mass movement of sediment on unstable slopes includes a continuous spectrum of processes according to the amount of deformation undergone by the sediment as it moves:

- *sliding* – displacement of the sediment mass along a slip plane, but accompanied by little internal deformation;
- *slumping* – displacement involving internal folding but not disaggregation of the sediment mass;
- *debris flowage* – movement of a mass of sediment as a slurry involving total re-organization and mixing of the sediment particles, ripping up clasts from the substrate;
- *turbidity flowage* – sediment carried in suspension, settling out to produce graded beds.

Subaquatic mass movement is considered to be a common process on the Antarctic continental shelf, which is relatively rugged and has slopes up to 15° (Wright & Anderson 1982). Unsorted ice-rafted debris may be transformed into sorted turbiditic sands in this way.

Mass movement is probably even more important on the Antarctic continental slope (although less steep), and may be intimately associated with rain-out of glacial debris when the grounding-line reaches the shelf break. There is also seismic evidence, in at least one place (Prydz Bay), to suggest that a considerable width (*c.* 10 km) and depth (*c.* 300 m) of the Antarctic shelf, probably composed of diamict, has suffered a rotational slide (Hambrey et al. 1991).

In comparison, ice-influenced Northern Hemisphere shelves are relatively subdued, and mass-movement processes are mainly limited to the continental slope, the troughs which extend across the shelf, and deltas bordering the coast. Continental slope slumping is well seen in seismic profile across the northwest British (Stoker 1990) and Barents Shelf margins (Boulton 1990, Vorren et al. 1990).

Scour by icebergs

The importance of iceberg scouring (or ploughing or gouging) on both northern and southern high-latitude shelves has recently been documented, following the deployment of geophysical techniques, such as side-scan sonar and bottom high resolution seismic profiling. Scour takes place when moving icebergs come into contact with the bed. In Antarctica many tabular icebergs have a draught of 500 m and cause considerable disruption of the sediment if they become grounded (Fig. 8.6). Northern Hemisphere icebergs are much smaller, but here the continental shelves are shallower. Iceberg scouring and subsurface deformation are potentially serious hazards to bed installations, such as hydrocarbon platforms, pipelines and cables.

Figure 8.6 Tabular icebergs stranded on the eastern Weddell Sea continental shelf, illustrating the extent to which sea-floor sediment may be subject to iceberg scour. Coats Land and the East Antarctic ice sheet are in the background.

Scour by sea ice

In contrast to icebergs, sea-ice scouring is confined to shallow water (Drewry 1986). Multi-year ice may reach an average thickness of 3 m, but the inverse sides of pressure ridges can scour the sea bed to depths of 30 m.

Bioturbation

Reworking of sediment by benthic organisms is common in both proximal and distal glaciomarine sediments. For example, homogenization of once-stratified diamicton or mudstone may be a major process, and can create confusion concerning interpretation of the depositional process. Bioturbation commonly is indicated by mottling of the sediment, and by distinct burrows. Much of the structureless sandy mudstone and diamictite in cores from McMurdo Sound has been interpreted as once-stratified sediment that has been bioturbated; in this case some shelly fossils of the organisms that caused the bioturbation are visible.

Compaction and recycling of sediment by ice re-advances

As a grounded ice mass advances across the continental shelf, the sediment becomes overpressurized. Water may be lost and the sediment becomes stiff (geotechnically **overconsolidated**), i.e. having a greater shear strength and bulk density than might be expected for its depth. Many polar shelf areas are covered by overconsolidated Pleistocene diamicts, indicating deposition from grounded ice, for example, the Barents Shelf (Solheim et al. 1990) and Antarctica (Elverhøi 1984). Sometimes, it is previously deposited glaciomarine sediment that has been loaded. In thick glacial sequences, like that in Prydz Bay, Antarctica, several overconsolidated horizons of diamictite are underlain by geotechnically and seismically defined unconformities.

Each new ice advance is potentially capable of incorporating older material. In particular, recycling of biogenic material may result in incorrect biostratigraphic sediment zonation and correlation of the sedimentary sequences. However, on a timespan of tens of millions of years this does not appear to have been a significant problem for dating either the McMurdo Sound or Prydz Bay cores.

8.4 Physical, chemical and mineralogical characteristics

8.4.1 Sediment texture

Sediment texture is usually expressed in terms of grain-size distribution, a parameter which in glaciomarine environments depends on:

(a) the size distribution of the source sediment, such as in basal, englacial or supraglacial debris, or on subaquatic glaciofluvial discharge or sedimentary gravity flowage;
(b) the energy level of the depositional environment as demonstrated by wave-grading and seaward fining of shallow marine settings.

Several studies have been made of the sand and finer grain-size distribution on high-latitude continental shelves, and a selection of distributions for a variety of facies are illustrated in Figure 8.7. The poorly sorted nature of most of these facies is a reflection of the source area, and a common origin from basal debris-rich ice for most of the sandy mud and muddy sand facies. In the drill-core CIROS–1, reworking processes are indicated by the massive sand facies,

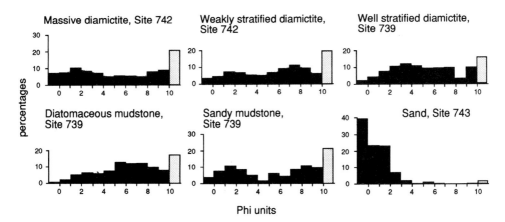

Figure 8.7 Grain-size distributions for glacigenic sediments recovered from the continental shelf of Antarctica, during Leg 119 of the Ocean Drilling Program to Prydz Bay, East Antarctica (extracts from Hambrey et al. 1991).

which is a sedimentary gravity-flow. Similarly, massive sand from the continental slope of Prydz Bay represents a deposit that has been reworked by currents.

Investigations by Anderson et al. (1980) have shown that plots of mean grain size versus sorting allow one to discriminate between basal tills deposited by grounded ice on the shelf, glaciomarine sediments derived from ice-rafting, and winnowed glaciomarine sediments (Fig. 8.8).

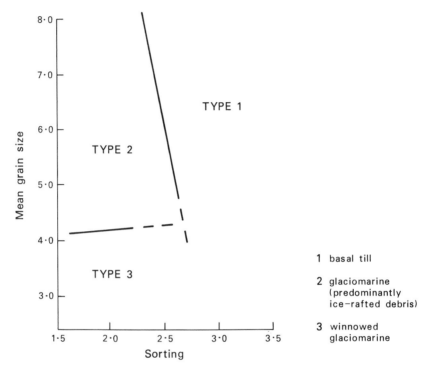

Figure 8.8 Fields illustrating mean grain size versus sorting for representative types of Antarctic glaciomarine sediment from the Ross, Weddell and Bellingshausen seas and George V Coast (summarized from Anderson et al. 1980).

Compared with unmodified glaciomarine sediments so far collected from Antarctica, diamicts on the Alaskan shelf (Eyles & Lagoe 1990) are texturally similar, although there is a suggestion that the clay component is more prominent. It is useful to compare the compositions of glacigenic samples from different areas, using ternary diagrams (triangular plots), in order to compare the thermal regimes of the glaciers that supplied sediment to the glaciomarine environment (Barrett 1986, Hambrey et al. 1991). Figure 1.3 demonstrates that Prydz Bay diamicts fall approximately between typical middle-latitude Pleistocene tills (Denmark) and present clay deposits from polar glaciers in Antarctica. A higher proportion of fines suggests that meltwater is an important component of the system. The Prydz Bay, and more especially the CIROS–1 samples, suggest that meltwater has played an important part during deposition of these predominantly Oligocene sediments. However, care must be taken in interpreting such data, as other factors, for example, changing bedrock source material, may also influence the grain-size distribution of the sediment.

8.4.2 Clast shape

The shapes of gravel clasts in glaciomarine sediments are inherited from those in the transporting medium (cf. Fig. 1.4). Glaciomarine sediments do not show the modified character that arises during deposition by lodgement. Thus, the basal debris that reaches the sea contains a mixture of rounded to angular clasts with strong modes in the subangular and subrounded categories, suggesting basal ice transport. However, some more southerly parts of Antarctica have a high proportion of angular and subangular clasts which, in the absence of any potential supraglacial source, may indicate erosion by subzero ice rather than by sliding (Fig. 8.9).

No systematic differences in clast shape are evident between ice-proximal and ice-distal settings, nor between different glaciological settings (ice shelves, grounded ice walls). Neither does there appear to be any lithological control on the shape of clasts; a comparison of

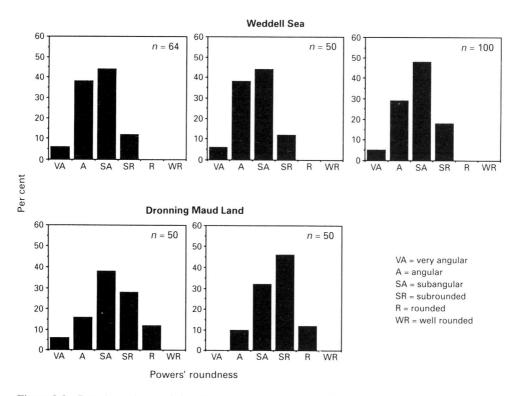

Figure 8.9 Roundness characteristics of gravel clasts in sediments from the continental shelf of Antarctica and the deep sea. Note the more angular nature of material from the Weddell Sea compared with that from the shelf off Dronning Maud Land. In the absence of any likely supraglacial source, the angularity may be related to fracture of bedrock by ice, accompanied by little, if any, basal sliding.

quartzite, granite, schist and metavolcanic rocks shows that each lithology has similar round-ness–sphericity distributions (Kuhn et al. 1993).

8.4.3 Surface features on gravel clasts

Stones carried at the base of a sliding glacier acquire a variety of surface features through contact with the bed. The most common feature in glaciomarine sediments is faceting, which, at its best developed, is evident in the form of flat-iron shapes (two parallel facets). Other clasts may be modified into bullet shapes. The important diagnostic feature of glacial trans-port, striations, occur on a widely varying proportion of clasts. Domack (1982) recorded striae on 12% of clasts from piston cores off the George V Coast, Antarctica; Barrett (1975) observed that 10% of stones sampled at Deep Sea Drilling Project sites in the Ross Sea were striated; and Hall (1989) recorded that 60% of stones extracted from the CIROS–1 core in McMurdo Sound. In contrast, the Prydz Bay cores yielded only a handful of striated clasts despite over 1000 m of drilling through glacigenic sediment (Barron et al. 1989). That surface markings are clearly dependent on lithology is indicated by data from the Lazarev Sea (Kuhn et al. 1993).

In Precambrian sediments in Svalbard, as many as 18% striated clasts have been extracted from a well bedded shaly diamictite of inferred glaciomarine origin (Dowdeswell et al. 1985), and significant percentages have been documented from many older sequences. The propor-tion of glacially transported striated clasts is mainly a function of lithology because in all these cases it is the fine-grained relatively soft lithologies that bear most striae, whereas coarse igneous and metamorphic clasts rarely carry striae.

8.4.4 Clast fabric

Few clast orientation studies have been made in contemporary glaciomarine environments because coring techniques normally do not allow orientation of the sample, nor do they yield enough pebbles for meaningful analyses to be undertaken.

Tests by the author on terrestrial sediments indicate that the orientation of sand-size grains is not significantly different from that of pebbles and large clasts. Thus, grain orientations can be used effectively for aiding the interpretation of a glacigenic sediment recovered from the sea bed. Ideally, three-dimensional measurements would be preferred to allow thorough statistical analysis, but this is rarely possible, and most measurements to date, apart from those undertaken by Domack & Lawson (1985), have been two dimensional.

Figure 8.10 illustrates two-dimensional data plotted as rose diagrams from massive and weakly stratified diamictites in the 702 m long CIROS–1 drill-core in the Antarctic. The data are also amenable to chi-squared statistical treatment in order to assess whether a preferred orientation is apparent at different significance levels. Massive diamictites interpreted as lodge-ment till deposited on the sediment floor show a strong preferred orientation, as do their ter-restrial counterparts. Other massive diamictites considered to be waterlain tills have a random fabric, as do most stratified diamictites, as would be expected from settling through water.

(a) Massive diamictites

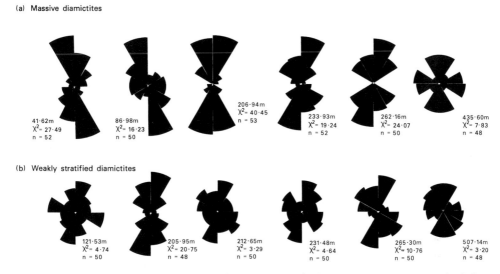

(b) Weakly stratified diamictites

Figure 8.10 Typical fabrics plotted as rose diagrams of sand-sized and larger material, from diamictites in the Oligocene–Miocene CIROS–1 drill-core, McMurdo Sound, Antarctica. Boxed figures include depth below sea floor, chi-squared values (significant at the 95% level for values greater than 11) and number of measurements; the latter is a measure of strength of preferred orientations. The first row illustrates massive diamictites of lodgement till (significant preferred orientation) and waterlain till (no preferred orientation). The second row is for weakly stratified diamictites deposited from floating ice (extracts from Hambrey 1989).

Random waterlain fabrics may be modified by slumping or currents, but these processes have not yet been fully investigated.

Obtaining samples from weathered pre-Pleistocene glaciomarine sequences exposed on land has proved to be a better prospect for obtaining clast fabrics, especially in three dimensions. Such studies confirm the patterns obtained for grains in more recent glaciomarine environments.

8.4.5 Geochemistry

The few systematic geochemical investigations of glacial sediments that have been undertaken demonstrate somewhat mixed success in interpreting palaeo-environments. The difficulty lies in interpreting analyses of sediments in which there are several genetically distinct components (Fairchild et al. 1988). Nevertheless, the potential value of whole-rock geochemistry has been demonstrated by various authors (e.g. Frakes 1975, Nesbitt & Young 1982, Fairchild & Spiro 1990)·with respect to the isotopic composition of carbonates in glacial sediments.

Major and trace-element geochemistry
Environments of deposition of terrestrial and marine glacigenic deposits are sufficiently varied to lead to differences in sedimentary geochemistry. Although glacial regimes are not usually characterized by active chemical weathering because of the low temperatures, the

proportions of dissolved elements are substantial – rock-flour particles are easily attacked because they have large surface areas. On land, oxygenated sediments dominate, whereas subaquatic sediments are more reducing and frequently strongly deficient in oxygen. Thus, one would expect elements that are easily precipitated in the presence of oxygen (e.g. iron and manganese) to be preferentially concentrated in terrestrial tills, but depleted in waterlain tills and glaciomarine sediments. This is borne out by the limited amount of data available. Norwegian terrestrial tills have iron abundances of 5–11%, whereas in Antarctic glaciomarine sediments they are 2–5%.

In contrast, some metals, such as Cu, Pb, Zn and V tend to form solids more readily under reducing conditions, as in the glaciomarine environment.

Biogenically influenced chemical sedimentation
Following ice-recession, the rôle of chemical sedimentation varies considerably. For example, in the Late Palaeozoic intracratonic basins of South Africa and Brazil, dark organic and siliceous shales accumulated, as well as in the Precambrian Adelaide geosyncline. Other postglacial sequences show regionally extensive pink, orange, cream dolostones with, or followed by, stratiform baryte and chert (e.g. the Neoproterozoic of Kimberley in Australia, the Grand and Petit Conglomerate in Zaïre, and in Mauritania and Svalbard).

The accumulation of siliceous shales and chert may reflect concentrations of silicon in basin waters as a result of dissolution of this element from silicate minerals in glacial meltwaters. This process is often related to acid volcanism, for example, in the Neoproterozoic Tindir Group of Alaska, in which thick chemical sediments are associated with glacigenic sediments. The silicon may be released when waters become sufficiently acid or alternatively removed from solution by siliceous microorganisms.

Stratiform baryte deposits are often associated with sedimentary phosphates and may therefore form by precipitation from sea water, such as during a transgression.

Carbonate isotope geochemistry and cementation
Until recently, most carbonates were interpreted as warm-water deposits, a view that led some authors (e.g. Schermerhorn 1974) to dismiss many supposed Precambrian tillites as non-glacial on the grounds that diamictites associated with carbonates or with a carbonate matrix could not be glacial. However, cold-water carbonates have been found in some modern polar marine settings. For example, in the Barents Sea clastic carbonates rest on glacigenic sediments (Bjørlykke et al. 1978) and in the Weddell Sea bioclastic carbonate sediments form an integral part of glaciomarine sediment (Elverhøi 1984). Much of the sea floor bordering the East Antarctic ice sheet in the eastern Weddell and Lazarev seas was found, during a cruise of the German icebreaker FS *Polarstern* in 1991, to be supporting well developed bryozoan colonies; often these were anchored on ice-rafted debris. However, the analogy made between bioclastic carbonate accumulations in the Barents Sea and carbonates associated with Proterozoic glaciomarine sequences is not necessarily a valid one, because skeletal organisms can precipitate calcite from waters undersaturated for carbonate, whereas cyanobacteria cannot (Fairchild & Spiro 1990).

The suggestion that carbonate precipitation from marine waters could occur when sea ice formed and increased the salinity, and the water subsequently warmed (Carey & Ahmed 1961) has not been substantiated. Nevertheless, a small percentage of carbonates occur in many Antarctic glaciomarine sediments today and probably plays a rôle in cementation. Cementation by carbonate is a major feature of the Oligocene CIROS–1 core in McMurdo Sound (Bridle & Robinson 1989), and also occurs in diffuse horizons in the contemporaneous Prydz Bay sequence (Barron et al. 1989), even though there are no obvious detrital or biogenic sources for the carbonate. The presence of extensive carbonates in Proterozoic glaciomarine sequences may suggest that the waters were oversaturated, but this hypothesis remains to be tested. The use of carbonate isotopes ^{18}O and ^{13}C in interpreting the palaeo-environmental significance of glaciomarine sediments is a promising new line of investigation that has already proved its value in interpreting Quaternary terrestrial tills and pre-Pleistocene glaciolacustrine sediments (Fairchild & Spiro 1990).

8.4.6 Mineralogy

Determination of the mineral fraction of the silt and larger fraction of glaciomarine sediment can assist in establishing its provenance, as it can for terrestrial sediments. Minerals of this size are not substantially altered in the marine environment before they are buried. However, new minerals may form, especially in deep ocean basins influenced by ice-rafting where sedimentation rates are low, e.g. manganese oxide coatings have been found on ice-rafted clasts in the southern Indian Ocean (Barron et al. 1989) and in the Weddell Sea during the 1991 cruise of FS *Polarstern*.

Clay minerology of glaciomarine sediments has been investigated in a variety of continental shelf and deep-sea areas, and provides important clues concerning the climate that prevailed on land before and during the input of glacial sediment to the marine environment. For example, various Antarctic studies (e.g. Anderson et al. (1980) in the Ross Sea, Elverhøi & Roaldset (1983) in the Weddell Sea, and Hambrey et al. (1991) in Prydz Bay) have indicated that illite, chlorite and smectite are the most common clay minerals, the first two of which are considered typical of marine sediments in high latitudes. Kaolinite is generally the product of warm, moist environments, but was found in large quantities in the early Oligocene part of the glacigenic sequence drilled during Leg 119 of the Ocean Drilling Program in Prydz Bay, reflecting erosion of a kaolinite-rich source rock and deep chemical weathering of the rock prior to glacial transport. The same signal of kaolinite input to the cold waters of the southern Indian Ocean was identified in well dated cores taken from the Kerguelen submarine plateau (Ehrmann 1991) and the comparison with Prydz Bay provides clear evidence of the onset of large-scale glaciation on East Antarctica.

8.5 Erosional features on the continental shelf and slope

Various erosional phenomena, mostly associated with grounded ice or subglacial meltwater, may be found on continental shelves at the present day. Some are similar to those occurring on land, but others are unique.

8.5.1 Submarine troughs

Submarine troughs are common bordering areas that have, or are being subject to, intense glacial erosion. Narrow troughs are continuations of fjords and essentially are genetically the same. Good examples occur in the northwest British, Norwegian and Greenland continental shelves. Broad troughs with gently sloping sides on continental shelves have also been interpreted as the product of glacial erosion, well developed examples include the Crary Trough (also known as the Filchner Depression) in the Weddell Sea, Antarctica which is over 1100 m

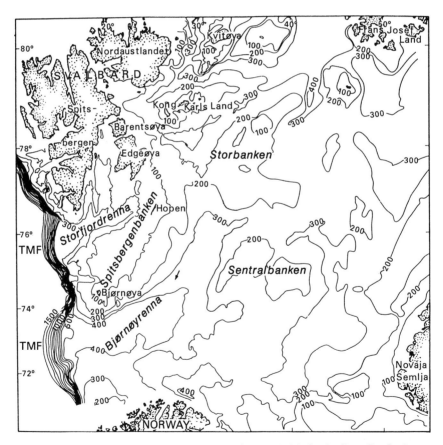

Figure 8.11 Bathymetry of the Barents Shelf, showing troughs/glaciated valleys (Storfjordrenna and Bjørnøyrenna) and trough-mouth fans (TMF) at the western edge of the shelf (from Solheim et al. 1990, with permission of Geological Society Publishing House, Bath).

242

deep in places, up to 200 km wide and extends across the continental shelf for 400 km beyond the present Filchner Ice Shelf limit (as well as a further 600 km beneath the ice shelf). It is closed by a sill at a depth of about 600 km (Fütterer & Melles 1990). On the Barents Shelf, Bjørnøyrenna deepens from 200 to 500 m towards the shelf edge through a distance of 600 km; its width is 200 km or more (Fig. 8.11).

Troughs, many tens of metres wide, and tens of metres deep, detected in high-resolution seismic records in the Ross Sea (Anderson & Bartek 1992), associated with major erosional surfaces, have also been interpreted as the product of glacial erosion. They are similar in scale to modern troughs and have been carved by ice streams. Sometimes, an **ice-stream boundary ridge** is developed between postulated ice streams, giving rise to a system of ridges and troughs; these may also be accretionary features.

Valleys cutting into the continental slope are also a feature of the Gulf of Alaska coast (Carlson et al. 1990). They are often continuations of fjords and continental shelf troughs. These valleys are erosional features, incised into the lithified strata that underlies the shelf. They have a U-shaped cross section and concave longitudinal cross sections that commonly shoal at their seaward limit. Diamicton drapes the walls of the valleys and the upper continental slope beyond. The shelf-valleys end abruptly at the shelf edge.

8.5.2 Tunnel valleys

U-shaped channels, measuring some 3 km wide and 100 m deep have been detected in high-resolution seismic records from the Ross Sea continental shelf, Antarctica (Anderson & Bartek 1992). These have been interpreted as large-scale subglacial meltwater channels, analogous to tunnel valleys on land. Similar features have been identified on the Scotian Shelf off Canada (Boyd et al. 1988) and in the Irish Sea (Eyles & McCabe 1989)..

8.5.3 Iceberg and sea-ice scours

The processes that create major erosional features on continental shelves influenced by iceberg drift are discussed in §8.3.4. Icebergs have variable keel geometries, and the scours reflect this geometry. Most common are multiple, parallel scour channels bordered by levées. In the Northern Hemisphere, iceberg scouring occurs as far south as the Labrador Shelf and Grand Banks off Newfoundland. Here, scours reach dimensions of 20–100 m in width and 2–10 m in depth, and one extending for more than 60 km has been reported. Both recent and relict iceberg scours have been described from the shallower parts of the Barents Shelf (Solheim 1991), where they have an average relief of 2–5 m and widths of 20–80 m (Fig. 8.12). Relict scours also occur in shelf areas that were affected by Pleistocene ice masses west of the British Isles and Norway.

Wherever the Antarctic shelf has been investigated, many scour marks up to 25 m deep and 250 m wide, are visible (Elverhøi 1984, Lien et al. 1989). It is difficult to imagine the deformation that might occur when mega-icebergs (100 km long) become grounded on shal-

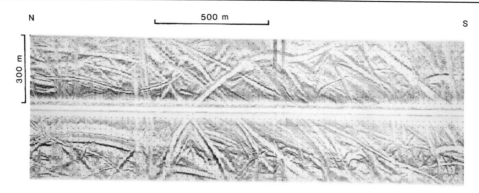

Figure 8.12 Iceberg ploughmarks on the floor of the Barents Sea in the vicinity of Bråsvellbreen, Nordaustlandet revealed by side-scan sonar techniques (photograph courtesy of Anders Solheim).

low banks. Not surprisingly, Elverhøi (1984) maintained that, in areas of present and former scour activity, the detailed stratigraphy of glaciomarine sequences may be disrupted.

Apart from linear scours, pits formed by wallowing of icebergs and by iceberg turbates have been observed from the Grand Banks off Newfoundland. Pleistocene iceberg scour marks have occasionally been reported (although mainly from lakes), but there are very few references to them in the older geological literature, yet they must be present in continental shelf sequences. Woodworth-Lynas (1990) has examined contemporary and Pleistocene scours in three dimensions, in order to assess the nature and depth to which deformation takes place beneath the keel.

Various names have been applied to sea-ice scours, including plough marks, scores, grooves and furrows. Sprag or jigger marks are formed by uneven movement across the sea floor (reviewed by Drewry 1986). Sea-ice scours have a rather different regional distribution compared with that of iceberg scours, and in the Arctic are much more widespread than iceberg scours. Although iceberg and sea-ice scours are morphologically similar, the latter tend to be of much lower relief and with finer, more closely spaced grooves. In Antarctica sea-ice scours are rare because of the deep water.

8.5.4 Slope valleys

Slope valleys are groups of gullies forming a dendritic pattern on the continental slope, and have been described from the Gulf of Alaska (Carlson et al. 1990). The smoothness of the valley sides suggests that there is a cover of sediment, dumped there when the glacier tongues occupied the shelf valleys. Gullies merge to form small canyons on the lower slope and small submarine fans occur at the outlets of some canyons. These gullies are believed to have formed as a result of erosion by sediment gravity-flows derived from terminal moraines directly at the shelf edge or directly from the glacier face as it was attacked by waves. The dendritic pattern on the Gulf of Alaska slope has been modified by tectonic compression on the rising continental margin.

8.5.5 Boulder pavements

Striated **boulder pavements** may form either at the base of grounded ice sheets or on inter-tidal flats where sea-ice abrasion takes place. In glacial settings, boulder pavements result from (a) the progressive accretion of boulders around an obstacle, (b) subglacial erosion of older sediment, or (c) by development of a lag deposit in a marine setting. It is the last process that concerns us here.

Boulder lag deposits on relatively shallow continental shelf areas are the product of wave and tidal-current winnowing of diamicton, especially when the sea level has been lowered. They become abraded by glacier ice as it advances across the shelf (Fig. 8.13). In the most comprehensive investigation of submarine boulder pavements, undertaken by Eyles (1988a) on the early Pleistocene Yakataga Formation in the Gulf of Alaska, it was envisaged that boulder-bearing diamict was deposited by rain-out of debris from icebergs or from an ice shelf. Advance of an ice shelf, partially in contact with the bed, resulted in the boulders acquiring flat, striated tops, but no significant reworking of the bed took place. Rising sea level led to rapid ice-shelf recession and an immediate resumption of diamict accumulation.

In contrast to shallow-marine boulder pavements, those formed on land and associated with lodgement till deposition show preferred orientation of boulders and typical bullet-nosed shapes, as well as a more varied facies association.

Boulder pavements have occasionally been reported in the older geological record, e.g. in

Figure 8.13 Striated boulder pavement formed by glacier "touch-down" on to a winnowed surface of rain-out diamicton, Yakataga Formation, Middleton Island, Gulf of Alaska.

the Late Palaeozoic glacigenic sequence in Brazil and the Neoproterozoic sequences of Greenland and Svalbard. Some of these, no doubt, could be reinterpreted in the light of the Eyles (1988a) model.

8.6 Depositional features on the continental shelf and slope

The application of geophysical techniques, including high-resolution and multi-channel seismic profiling, and side-scan sonar, has expanded our knowledge of the depositional forms on ice-influenced continental shelves in the past decade. Some have been interpreted in the light of supposedly analogous features on land, but other forms have no direct counterparts on land.

8.6.1 Fluviodeltaic complexes on the continental margin

Along the Gulf of Alaska, in Iceland, Greenland and Svalbard proglacial fan deltas topped by sandur plains and building out from the coast in proximity to piedmont glaciers occur. The largest are associated with the Bering and Malaspina glaciers in Alaska. The plains are characterized by gravelly longitudinal bars with cross-bedded sands and gravels. The coastline areas are dominated by waves, and river-mouth bars and spits extend several thousand metres in the direction of the prevailing onshore drift. Lagoons often occur behind the beach ridges, and sometimes are directly influenced by the ice. Prodelta sediments are sandy to depths of about 50 m and up to 15 km offshore, and pass into clayey silts further out on to the continental shelf.

Coastal fluviodeltaic complexes forming more distally from the source glaciers tend to have more stable and abandoned parts, allowing marshes to develop with organic muds, and channels to be filled by the products of estuarine processes. One example is the Copper River in southern Alaska, which probably has the largest delta dominated by glacial meltwater in the world. The braided delta top is subjected to aeolian processes, and large sand dunes have developed in places. The shoreface is complex comprising, in a seaward direction, marginal islands, a breaker bar, middle shoreface sands, lower shoreface sands and muds, and prodelta to shelf muds.

8.6.2 Delta-fan complexes

Detected by high-resolution seismic profiling, a **delta-fan complex** is the result of deposition beneath, and close to, the grounding-line of an ice sheet a long way from the open sea (Anderson & Bartek 1992). The complex is essentially a prograding sequence with prominent lamination illustrative of glaciomarine deposition, as well as gravity-flow material derived from basal glacial debris. A good example, several hundred metres thick, occurs on the outer shelf of the Ross Sea. Its upper surface is elevated compared with the level of continental shelf behind, thus forming a bank on which icebergs are prone to grounding. Delta-fan complexes have also been identified in the middle of the continental shelf.

8.6.3 Subglacial deltas

Subglacial deltas are also prograding features, but lack stratification in the proximal parts. They grade into laminated deposits in the delta bottom-set deposits, and down-lapping on to glacial erosional surfaces (Anderson & Bartek 1992). They are believed to form by "conveyor-belt" recycling of soft subglacial sediment and they provide a platform over which the ice may advance, according to the model of Alley et al. (1989). These features, too, have been identified in the Ross Sea.

8.6.4 Till tongues

During seismic surveys of the eastern Canadian and Norwegian continental shelves, King & Fader (1986) discovered wedge-shaped bodies characterized by acoustically indecipherable reflections, interfingering with stratified glaciomarine sediments, and connected to large off-shore moraines (Fig. 8.14). These authors inferred that the wedges comprised subglacially derived till, and they named the features **till tongues**. They inferred that the features formed at an oscillating ice margin where the ice became buoyant at the grounding-line. King et al. (1991) modified the till-tongue model to include a proglacial apron comprising a series of sediment gravity-flows. Till tongues have also been documented from seismic record in the

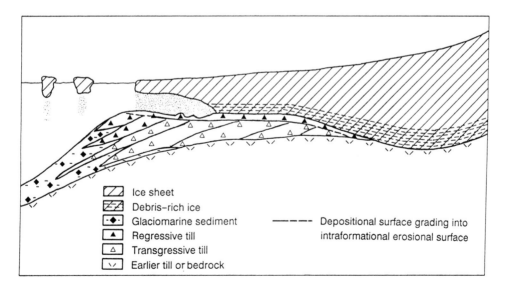

Figure 8.14 Schematic representation of till-tongue associations formed under both transgressive and regressive regimes (modifed from King et al. 1987).

247

Ross Sea, Antarctica, where they rest on erosion surfaces, thus providing a means of identifying glacial advances in seismograms (Anderson & Bartek 1992).

8.6.5 Trough-mouth fans

Trough-mouth fans are major features of continental margins, across which large ice streams drained in the past, and they consist of arcuate fans of prograding sediment. The large ones, such as the 400 km wide Bear Island (Bjørnøya) Trough-Mouth Fan on the Barents Sea margin (Fig. 8.11) and the 300 km wide Prydz Trough-Mouth Fan in East Antarctica, reflect sedimentation from major marine-based ice streams at the grounding-line or at the calving limit when they reached the continental shelf break. Little is known of the facies in trough-mouth fans, although drilling through prograding sediments down to 100 m below the sea floor at the mouth of Prydz Bay yielded a sequence of unconsolidated massive diamicton. This diamicton was believed to represent advanced ice, the sand indicating reworked sediment deposited during interglacials (Barron et al. 1989). Seismic records indicate that slumping occurs frequently.

The sediments on the top of a smaller trough-mouth fan, at the mouth of Isfjordrenna Spitsbergen, comprise an upper part of proximal glaciomarine mud capped by a gravel lag, and a lower part with dense, massive unfossiliferous diamicton deposited by grounded ice (Boulton 1990). Frequent slope failure is evident here also, but mainly as small-scale mass movements. The accretion of trough-mouth fans is aided by erosion and shelf-over-deepening on the inner parts of the continental shelf.

8.6.6 Diamict(on/ite) aprons

Diamict(on/ite) aprons are features (named by Alley et al. 1989 as till deltas) that have been inferred, rather than directly observed, to form at the grounding-line of ice shelves, floating glacier tongues or ice margins grounded to the edge of a steep slope, such as the continental shelf break. The processes involved in their formation have been described earlier (§8.3.1). They comprise diamicton released at the break in slope by a combination of shearing of saturated till at the base of the ice, rain-out at the grounding-line and slumping of this sediment down the slope. They develop not from point sources but from a continuous line along the break in slope. They may form the bulk of the material in trough-mouth fans, but are not confined to them. The prograding sequence in Prydz Bay, East Antarctica, is a good example (§8.9.1). Diamict aprons are forming extensively today only at the grounding-lines of ice shelves and floating glacier tongues in Antarctica, well back from the continental shelf break.

8.6.7 Shelf moraines

It has long been known that prominent ridges at the edge and in the middle of high-latitude continental shelves are probably end-moraine complexes. King et al. (1987, 1991) have investigated such features created by the Fennoscandian ice sheet using seismic techniques on the mid-Norwegian continental shelf. They identified three types of moraine complex.

- **Linear moraines** are composite features consisting of (a) a lower stacked till-tongue succession, resulting from subglacial deposition near the grounding-line under conditions of advance, and (b) an unconformably overlying upper part formed during recession. Linear moraines are distinct ridges running for hundreds of kilometres at the shelf edge.
- **Tabular moraines** form in an intermediate-shelf position. They are irregular in plan view, with lateral dimensions as great as 70 km and a relief of 50–75 m. They have both abrupt and diffuse boundaries with the surrounding topography. On Trænabanken there are as many as five such complexes inside the linear moraine.
- **Hummocky moraines** are less extensive, but have a relief that may be more pronounced (25–100 m).

Both tabular and hummocky moraines are thought to have a common origin, formed where active ice was thinning and becoming buoyant over a broad zone in the form of a grounding zone, rather than a grounding-line. Deposition may be enhanced by a simultaneous rise in sea level. Both these types of ridge were prone to iceberg scour during the period of recession of the ice sheet.

Submarine end-moraine complexes have also been identified in seismic profiles on the northern Hebrides and West Shetland continental shelves (Stoker & Holmes 1991).

8.6.8 Flutes and transverse ridges

Largely depositional forms, resulting from a grounded ice mass sliding on, or deforming, the sea bed on the northwestern Barents Sea, have been investigated using geophysical methods. They are represented by a system of parallel grooves and ridges, comprising stiff diamicton (Fig. 8.15). In several places the groove-and-ridge system is associated with short straight-to-arcuate ridges that run approximately perpendicular to the strongly linear pattern (Fig. 8.15). By comparision with terrestrial depositional landforms, the grooves and ridges are interpreted as a set of **flutes**, described in §4.5.2.

The flutes are often 100–500 m long, but lengths over 1 km have been recorded. Their widths range from 1 to 15 m, with 4–8 m being typical. The fluted sea floor covers an area of around 4000 km^2, in water depths of between 160 and 300 m, in northernmost Bjørnøyrenna (Fig. 8.11). They are parallel to the inferred ice-flow direction down the submarine trough.

The flutes run across the associated transverse ridges, and through the gaps between ridge segments. The transverse ridges vary in length from 100 to 500 m, and in width from 15 to 30 m. They reach heights of 8 m. The ridges are intepreted as De Geer moraines, formed by seasonal ice-push at, or close to, the grounding-line.

In the special case of marine-based surging glaciers, Solheim (1991) described from Bråsvellbreen, Nordaustlandet, Svalbard, a distinct suite of sea-floor morphologies, associated first with rapid advance and then with stagnation of strongly fractured ice. Many discontinuous arcuate ridges, which formed subparallel to the ice margin during the surge, extend

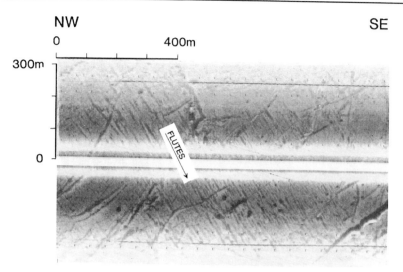

Figure 8.15 Flutes and superimposed transverse ridges (De Geer moraines) on the floor of the Barents Shelf revealed by side-scan sonar (photograph courtesy of Anders Solheim).

for several hundred metres across the surge zone (Fig. 8.16). A set of linear ridges forming a rhombohedral pattern is also developed.

8.6.9 Mass-movement features

Oversteepened slopes with poorly consolidated diamicton and glaciomarine sediment are unstable, so submarine slides, slumps and gravity-flows are common in the glaciomarine

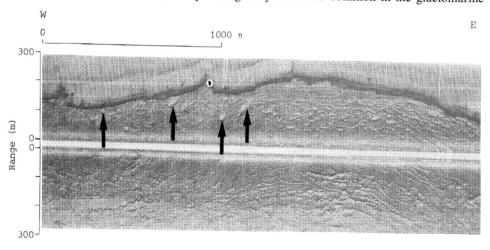

Figure 8.16 Side-scan sonar image of discontinuous, arcuate ridges, formed subparallel to the front of surge-type Bråsvellbreen, northern Barents Sea. Possible iceberg-impact features close to the ice margin are arrowed (photograph courtesy of Anders Solheim).

environment, as discussed in various contexts already. These features have a hummocky, sometimes lobate, morphology, frequently with a relief of several metres and extend laterally for hundreds of metres. Larger collapse features on a scale of tens of kilometres occur occasionally at glaciated continental margins.

8.6.10 Raised beaches

These are common features of glacio-isostatically uplifted areas bordering continental shelves. Raised beaches consist of well sorted sand and gravel with rounded clasts, originally supplied mainly from glacial or fluvioglacial sources. The best preserved raised beaches are of Holocene age, and good examples occur in northwest Europe, Canada, the USA, and the sub-Antarctic islands. Pre-Holocene raised beaches have commonly been modified or obliterated by later glacial erosion.

8.7 Sediment distribution patterns

A brief outline is given here of the areal distribution of glacigenic sediment in some continental shelf settings that today are directly or indirectly under the influence of ice.

8.7.1 Seawards of Antarctic ice shelves

Weddell Sea
Sediment patterns in the Weddell Sea are variable and related to bathymetry and current activity (Fütterer & Melles 1990). In front of the Filchner–Ronne Ice Shelf, the grain-size distribution of surface sediments is related to water depth and is interpreted to be mainly current-controlled (Haase 1986). Very well sorted fine to medium sand and moderately well rounded pure sand occur to the west of Gould Bay in water depths of about 250 m. The high current velocities (up to 20 cm sec^{-1}) required to produce such sediments are tidal (Haase 1986), as confirmed by one current measurement of 40 cm sec^{-1} perpendicular to the ice front (Robin 1983). As water depth increases towards the west, the mud fraction increases, and therefore slower currents are inferred. In addition, the increase in the gravel component towards the Antarctic Peninsula illustrates the enhanced rôle of iceberg-rafting. A higher proportion of sand on the eastern flank of the Ronne Trough is related to a deep current flowing south under the ice shelf. Sub-bottom seismic profiling north of the Ronne Ice Shelf often failed to penetrate the gravelly floor. Elsewhere, the sediments are interpreted as having been compacted by Pleistocene grounded ice.

In the Crary Trough, the surface sediments are mainly depth-dependent and they consist of pebbly and sandy muds of glaciomarine origin. The western flank of the trough shows a deepening trend from sand to mud along its axis. On the eastern flank the same fining-downwards trend is revealed, but the sediments are gravelly (diamictons) on the upper continental slope and near the Filchner Ice Shelf edge. Grain-size distribution is controlled by current circulation and input of ice-rafted debris. Sands indicate the effect of currents that winnow out the fine material. The surface sediments are unconsolidated and are mainly up to 1 m in thickness, underlain by overconsolidated diamicton (interpreted as a basal till of late Wisconsinan age). Unlike many other parts of the Antarctic margin, the biogenic component forms only a minor proportion of the total sediment.

Ross Sea

Biological productivity in the Ross Sea is much higher than in the Weddell Sea, and this is reflected in the proportion of silicon derived from diatoms in the sediments. Dunbar et al. (1985) have shown that the surface sediments in the Ross Sea are mixtures of unsorted ice-rafted debris, siliceous biogenic material (diatom fragments), calcareous shell debris, terrigenous mud transported in suspension. Terrigenous mud accounts for up to 50% of the surface sediment along the eastern part of the Ross Ice Shelf front, while to the west this is mixed with 10–50% of biogenic silica. An interesting feature of the ice-rafted debris distribution is that it increases in an offshore direction. The low concentration near the calving front suggests that most basal melt-out occurs near the grounding-line, while on the outer shelf debris-bearing icebergs from valley glaciers are more frequent.

8.7.2 Continental shelf bordering mountain terrain with temperate glaciers

The coast of British Columbia and Alaska today is a region characterized by a continental shelf flanked by high, tectonically active mountains, which still support many tidewater glaciers. Analogous settings are probably common in the older geological record, notably that of the Neoproterozoic age. The distribution of superficial sediments on the coast of the northern Gulf of Alaska (Carlson et al. 1990), seawards of the major fjords and glaciofluvial systems is illustrated in Figure 8.17. In the simplest case, a graded-shelf sedimentary profile appears to be developing today: sand accumulating in the nearshore zone grades outwards into clayey silt in mid-shelf, the principal source being sediment brought to the coast by glacial meltwater. This Holocene facies association wedges out near the outer shelf, where diamicts, consisting of gravelly sand, muddy gravel and pebbly mud crop out extensively; these diamicts are probably of Pleistocene age.

Tectonic activity in places has uplifted the banks (e.g. Tarr Bank, Middleton Island and Kayak Island, Fig. 8.17), periodically permitting the fine sediment to be winnowed from the diamict and creating cobble and boulder lag deposits, which have been subject to abrasion during a subsequent ice-advance (§8.5.5). The modern glacially formed submarine troughs or sea valleys have concentrations of relict coarse sediments adjacent to their mouths near the outer shelf, and may represent end-moraine complexes. These troughs are significant sedimentary traps, and seismic reflection profiles show accumulations of modern sediment tens to hundreds of metres thick.

The continental shelf edge and upper slope is covered mainly by gravelly sandy silts, or diamicts (with up to, or more than, 25% gravel) and silty sands. The sediment was probably

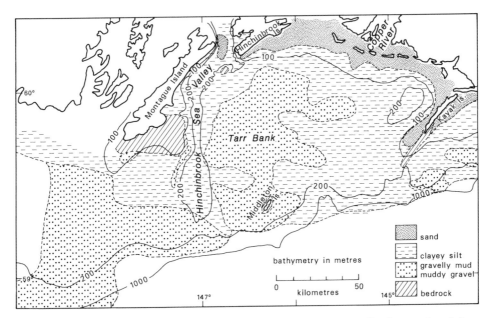

Figure 8.17 Sediment distribution map of the Gulf of Alaska continental shelf and upper slope (after Carlson et al. 1990, with permission of Geological Society Publishing House, Bath).

deposited at the shelf edge during one of the most recent glacial stages when sea level was comparatively low, either as glacigenic mud with dropstones or as a result of resedimentation of basal tills.

8.7.3 Continental shelf partly influenced by cold glaciers

The Barents Shelf is one of the broadest continental shelf areas in the world and during the Pleistocene Epoch it was covered by ice. Sedimentation today is influenced by cold glaciers on the rugged Svalbard Archipelago in the north, and much of the northern shelf is covered by sea ice in winter. The surface sediments of the western shelf have been mapped by the Norwegian Polar Research Institute (Elverhøi 1984).

Late Weichselian blue-grey glaciomarine sediments are overlain transitionally by olive-grey mud partly derived from suspended matter of glacial meltwater origin. In the north the mud is pebbly as a result of iceberg- and sea-ice-rafting. On Spitsbergenbanken biogenic material (shell fragments and barnacles) make up 60–90% of the surface sediments as a result of current winnowing. On the southern and southeastern slopes of Spitsbergenbanken the olive-grey mud is missing and a cobble lag, resulting from strong bottom-current activity, is present.

253

8.8 Sedimentation rates

Sedimentation rates on continental shelves are generally several orders of magnitude less than in fjords (cf. §7.5). In the polar regions, estimates for Holocene sedimentation include $0.02-0.05\,mm\,yr^{-1}$ for the Weddell Sea (Elverhøi & Roaldset 1983) while $0.02-0.07\,mm\,yr^{-1}$ is indicated for the Barents Sea (Elverhøi 1984).

Over longer time-intervals rates have been determined for the CIROS–1 core in McMurdo Sound. The lower, early Oligocene part of the core gives an average sediment rate of $0.2\,mm\,yr^{-1}$, whereas the upper part (mainly late Oligocene to early Miocene) is $0.04\,mm\,yr^{-1}$ (Harwood et al. 1989). The sharp difference may reflect a greater tendency to uplift of the neighbouring Transantarctic Mountains during the earlier phase. The longer term average sedimentation rate for progradation of the outer Prydz Bay continental shelf is $0.008-0.012\,mm\,yr^{-1}$, although large parts of the sequence are missing and may have been eroded.

In contrast to these polar continental shelves, sedimentation rates are at least an order of magnitude higher on the cool temperate Gulf of Alaska Shelf. From the thickness and age estimates given by Eyles & Lagoe (1990), an average sediment rate of $1\,mm\,yr^{-1}$ is obtained.

In the deep ocean basins, sedimentation tends to be much more continuous, but overall the rates tend to be of the same order of magnitude. For example, the Ocean Drilling Program recorded $0.013\,mm\,yr^{-1}$ at Site 645 in southern Baffin Bay and $0.03\,mm\,yr^{-1}$ at Site 745 in the southern Indian Ocean, much of the accumulation being terrigenous material of glacial derivation.

8.9 Sedimentary facies in glaciomarine environments

Although the range of processes operating in the modern glaciomarine environment has now been widely investigated (§8.3), little is known of the nature of sedimentary sequences accumulating in such environments (Eyles & Lagoe 1990). However, the older geological record provides excellent opportunities for studying thick sequences of glaciomarine sediment. The best exposed, and by far the thickest, Late Cenozoic glaciomarine sequence occurs along the Gulf of Alaska. Here, 5 km of glaciomarine strata belonging to the Yakataga Formation, have been progressively uplifted by tectonic collision of the Pacific Ocean and North American plates, and span 5–6 Ma (Eyles & Lagoe 1990). Even longer records (35–40 Ma) have been recovered from drill-holes on the Antarctic continental shelf, although these show some hiatuses (Barrett 1989b, Cooper et al. 1991b, Hambrey et al. 1992). Many glaciomarine successions of pre-Cenozoic age have also been documented. Neoproterozoic strata provide several excellent glaciomarine successions. Andrews & Matsch (1983) have provided a bibliography of papers describing glaciomarine sequences, and Anderson (1983) has summarized the main occurrences in the geological record, based on the global survey compiled by Hambrey & Harland (1981).

8.9.1 Sedimentary structures in glaciomarine sediments

Glaciomarine sediments demonstrate a wide range of sedimentary structures reflecting their composite origin. Sediment deposited subaquatically from ice by melt-out, without reworking, is a non-sorted, homogeneous diamicton, although over several metres there may be slight variations in the content of gravel (Fig. 8.18a).

As soon as these sediments are subject to bottom currents, mud may be preferentially winnowed out leaving rippled and scoured surfaces (Fig. 8.18b) giving rise to stratified diamicton (proximal glaciomarine sediment) with bedding typically on a centimetre-scale. Wispy bedding and signs of loading are common, reflecting the high water content (soupiness) of the sediments (Fig. 8.18c & d).

Bedding is often punctured by dropstones (Fig. 8.18e), over which subsequent sedimentary layers are draped. Many elongate dropstones stick in the sediment vertically, if it is not too soupy. Some dropstones occur as clasts of diamict or glacial mud (till pellets) and are assumed to have been dropped in as frozen lumps; usually they are up to a few centimetres in diameter, although some are bigger. Stronger currents remove sand as well as mud, leaving a gravel lag, which has a gradational relationship with the underlying undisturbed sediment but is followed sharply by later sediments. If sedimentation has been continuous, then diamictons often show gradational boundaries from stratified to non-stratified varieties as well as into less ice-influenced muddy sands and sandy muds. Successions may be tens of metres thick and may lack distinct bedding hiatuses.

Recycling of diamictons by mass-movement processes generates new sedimentary structures. Slumping of stratified diamicton gives rise to isoclinal folds. In slides, partial disaggregation of beds may produce rootless folds detached from the original bed. Submarine debris-flows commonly affect glacigenic sediments. They tend to be represented by beds ranging in thickness from a few centimetres to several metres (Fig. 8.18f). The end product may still be a massive diamicton, difficult to distinguish from the original material. However, the flow normally has a sharp top and bottom, contains clusters of several gravel clasts, ripped-up from the bed over which it passes, and occasional faint inverse or normal grading.

If mass movement of diamicton develops into a turbidity-flow, the end product shows normal grading, usually of the "coarse-tail" type (i.e. in which larger clasts may occur throughout the graded bed but in smaller quantities towards the top). Both debris-flows and turbidity-flows may load the underlying sediment, causing it to deform in a convolute manner or to penetrate the debris-flow as a sedimentary dyke. Sandy and muddy glaciomarine sediments of more distal origin are subject to the same reworking processes as diamictons; their structures are typical of continental margins generally, with the added influence of floating ice. The glaciomarine environment may yield rhythmically laminated sediments, although these are much less common than in fjords. Normally the laminae are graded, representing turbidity-current deposits, and they can range from proximal to distal.

Ice advancing on its bed across glaciomarine sediment, such as on a continental shelf, may abrade gravel lag deposits or accumulations of ice-rafted gravel to give a boulder pavement. If the underlying sediment is partially consolidated, it may fracture, and slickensides may form on the fault planes. More intensive glaciotectonic deformation may occur to depths of tens of

255

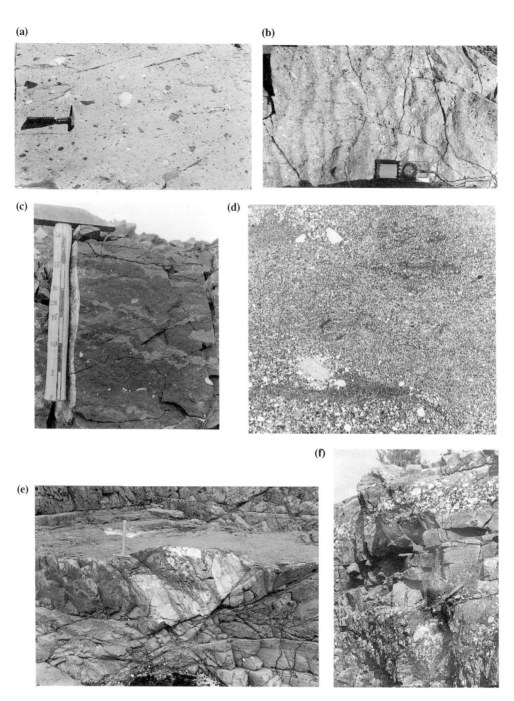

Figure 8.18   Features in sediments in the continental shelf glaciomarine environment. (a) Massive diamictite, interpreted as waterlain till, Neoproterozoic Ulvesø Formation, East Greenland. (b) Wave ripples and winnowed surface on top of diamictite bed, Neoproterozoic Wilsonbreen Formation, NE Spitsbergern. (c) Clast-poor diamictite with wispy, diffuse stratification, the result of slight winnowing of soupy waterlain till followed by slumping, Wilsonbreen Formation, NE Spitsbergen. (d) Thin-section of rock shown in (c), showing variable grain-size distribution, including clay-rich wispy beds (dark) and diffuse winnowed silty beds (bottom and top). (e) Large dropstone of pegmatite in sandy turbidites, top of the Dalradian succession at Macduff, Banffshire, Scotland (age: Ordovician or Neoproterozoic). (f) Matrix-supported conglomerate horizons interpreted as subaquatic debris-flows, interbedded with siltstone of glaciomarine origin, Whitefish Falls area, Ontario, Canada.

metres, manifested in thrusts and overturned folds and the incorporation of rafts of underlying sediment occasionally more than a hundred metres long. In this respect, these structures are similar to those of advancing ice on land.

8.9.2 Lithofacies in the glaciomarine environment

Lithofacies are normally described in terms of the following: diamicts, gravel, sand, mud and combinations thereof, and their lithified equivalents. Subfacies are designated according to whether the rock is massive, stratified, or bioturbated, or has other distinguishing features. These lithofacies are then interpreted in the context of the whole sequence and by using other parameters (§8.5). Massive diamict (Fig. 8.19a) can variously be interpreted as a lodgement till, deformable till beneath ice, waterlain till, or a debris-flow, or as a distal glaciomarine mud with ice-rafted debris that has been bioturbated or deposited in a quiet-water basinal setting. Weakly to well stratified diamict (Fig. 8.19b) represents the transition to sediment increasingly dominated by marine processes, notably reworking by currents in which the influence of ice-rafting becomes clearer (i.e. proximal glaciomarine). Muddy sands to sandy muds demonstrate the transition to distal glaciomarine with ice-rafted debris becoming relatively less important. Other facies represent the processes that rework the sediment after deposition, notably mass movement (Fig. 8.19c & d), downslope channelling (Fig. 8.19e), and by the addition of a biogenic component in the form of both macrofossils and microfossils such as diatoms (Fig. 8.19f).

Lithofacies have now been described from many Cenozoic and older sequences. The Antarctic continental shelf displays a wide range of facies, including sediments released directly from glacier ice, which are summarized and interpreted according to the proximity of ice and water depth in Table 8.2. Relative proportions have also been calculated from which it may be noted that diamictite, sandstone and mudstone predominate. Similar glaciomarine facies occur elsewhere, although of course the biogenic influence varies according to age and was negligible in the Proterozoic era.

The range of facies that accumulate on the continental slope during periods when an ice sheet is grounded to a continental shelf edge has been investigated on the Norwegian and Barents continental slopes using a combination of shallow coring and seismic techniques (Yoon et al. 1991). Both diamictons and mud lithofacies, containing abundant signs of downslope movement have been examined. These are listed in Table 8.3, together with their interpretations.

8.9.3 Glaciomarine facies associations

Facies associations are normally assemblages of various lithofacies in vertical sequences of strata that display a common theme, e.g. glaciomarine. As such they reflect environmental changes through time. However, it is necessary to consider how these facies change in an areal sense, since this will explain the vertical associations. Thus, a typical present-day transition from waterlain till to distal glaciomarine is best represented by a recession sequence in

257

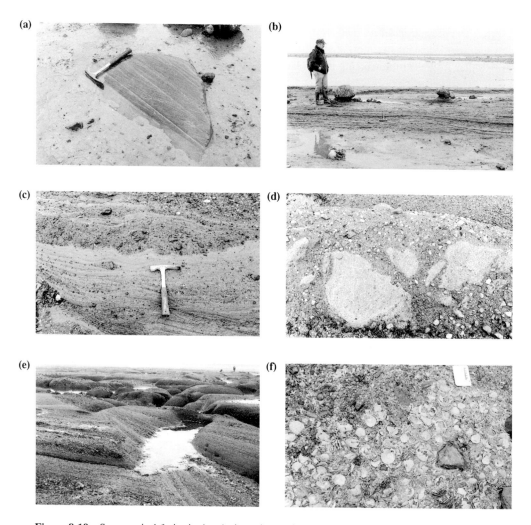

Figure 8.19 Some typical facies in the glaciomarine environment, Miocene-Pleistocene Yakataga Formation, Middleton Island, Gulf of Alaska. (a) Massive diamict with striated clast, resulting from rapid deposition of mud, together with ice-rafting. (b) Stratified diamict formed where some sorting by bottom currents of material as in (a) had taken place. (c) Stratified diamict with prominent gravel lag horizon. (d) Deformed masses of sand, the largest 0.5 m in diameter enclosed within diamict facies, resulting from dismemberment of beds during loading. (e) Large-scale cross-bedding in a sand and gravel member at base of the Yakataga Formation; these represent slope deposits, reworked from glacial material on the shelf. (f) Coquina bed, formed during a phase of little sedimentation, except for the dropstones, when current winnowing was taking place.

a vertical profile, with massive diamictons at the base passing up into mudstones with dispersed stones.

Figure 8.20 presents an idealized example of the type of facies association that may develop as a temperate glacier influences, and then advances across, a continental shelf, but many other associations may arise according to the thermal and hydrological characteristics of the glacier.

Table 8.2 Facies recovered in deep drill-cores taken from the Antarctic continental shelf, and their interpretation (summarized from Hambrey et al. 1991).

Facies	Description	Interpretation
Massive diamictite	Non-stratified muddy sandstone or sandy mudstone with 1–20% clasts; occasional shells and diatoms	Lodgement till (with preferred orientation of clast fabric) or waterlain till (random fabric)
Weakly stratified diamictite	As massive diamictite, but with wispy stratification; bioturbated and slumped; partly shelly and diatomaceous	Waterlain till to proximal glaciomarine sediment
Well stratified diamictite	As massive diamictite with discontinuous and contorted stratification; occasional dropstone structures; abundant diatoms and shells	Proximal glaciomarine/glaciolacustrine sediment
Massive sandstone	Non-stratified, moderately well sorted sandstone, with minor mud and gravel component; loaded bedding contacts	Nearshore to shoreface with minor ice-rafting in distal glaciomarine setting; better sorted sands with loaded contacts are gravity flows; associated with slumping
Weakly stratified sandstone	As massive sandstone, but with weak, contorted, irregular, discontinuous, wispy, lenticular stratification; brecciation, loaded contacts and bioturbation	Nearshore with minor ice-rafting in distal glaciomarine setting; better sorted sands are gracity flows; associated with slumping
Well stratified	As massive sandstone, but with clear often contorted stratification	Nearshore with minor ice-rafting in distal glaciomarine setting; some slumping
Massive mudstone	Non-stratified, poorly sorted sandy mudstone with dispersed gravel clasts; intraformational brecciation and bioturbation; dispersed shells and shell fragments	Offshore with minor ice-rafting in distal glaciomarine setting; some slumping or short distance debris flowage
Weakly stratified mudstone	As massive mudstone, but with weak, discontinuous, sometimes contorted stratification defined by sandier layers; bioturbated	Offshore to deeper nearshore with minor ice-rafting in distal glaciomarine setting; slumping common
Well stratified mudstone	As massive mudstone, but with discontinuous well defined stratification, with sandy laminae; syn-sedimentary deformation and minor bioturbation	Deeper nearshore with minor ice-rafting in distal glaciomarine setting; some slumping
Diatomaceous ooze/ diatomite	Weakly or non-stratified siliceous ooze with >60% diatoms; minor components include terrigenouss mud, sand and gravel	Offshore, with minor ice-rafting in distal glaciomarine setting
Diatomaceous mudstone	Massive mud or mudstone with >20% diatoms and minor sand	Offshore with sedimentation predominantly influenced by ice-rafting and underflows in distal glaciomarine setting
Bioturbated mudstone	As massive mudstone, but stratification highly contorted or almost totally destroyed by bioturbation	Offshore to deeper nearshore with minor ice-rafting; extensively burrowed
Mudstone breccia	Non-stratified to weakly stratified, very poorly sorted, sandy mudstone intraformational breccia with up to 70% clasts; syn-sedimentary deformation and minor bioturbation	Offshore to deeper nearshore slope-deposits with minor ice-rafted component, totally disrupted by debris flowage
Rhythmite	Graded alternations of poorly sorted muddy sand and sandy mud; stratification regular on a mm-scale; dispersed dropstones	Turbidity underflows derived from sub-glacial source, with ice-rafting in a proximal glaciomarine setting
Conglomerate	Non-stratified to weakly stratified, poorly sorted, clast to matrix supported sandy conglomerate; normal and reverse grading evident; clasts up to boulder size; intraclasts if mudstone frequently incorporated; loading and other soft-sediment features present	Slope debris-flows derived directly from proglacial glaciofluvial material, or from subaqueous discharge from glacier; well defined beds may be fluvial

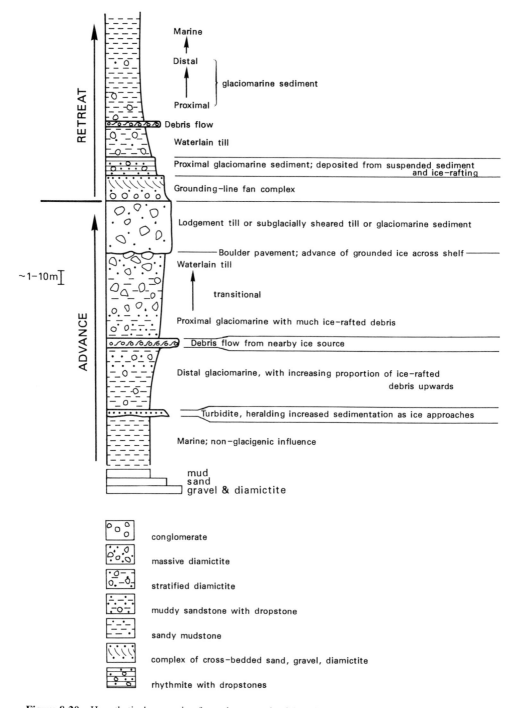

Figure 8.20 Hypothetical succession formed as a result of the advance and recession of a temperate glacier across the continental shelf.

Table 8.3 Facies recorded in shallow cores taken from the continental slope off northern Norway and the SW Barents Shelf (summarized from Yoon et al. 1991).

Facies	Characteristics	Depositional process
Thick-bedded disorganized mud	Coarse-grained clasts dispersed in fine matrix without internal organization; more than 1 m thick; bioturbation minimal	Debris-flowage
Thin-bedded disorganized mud-clast mud	Abundant mud clasts and rock fragments randomly scattered in mud matrix; bioturbation slight to common; variable thickness (1–10 cm)	Vertical settling of ice-rafted debris
Silt-clay couplet	Couplet composed of a basal slit unit and an overlying clay unit; basal silt thinly laminated and clay unit homogeneous; individual unit < 5 cm thick; bioturbation restricted to upper part; facies boundaries well defined	Fine-grained turbidity current
Indistinctly laminated mud	Poorly sorted mud showing irregular and discontinuous laminae a few decimetres thick; bioturbation minimal	Downslope bottom current with high sediment fallout rate
Layered mud	Irregular alternation of thin (a few mm) silt-rich and silt-depleted mud layers; bioturbation slight to common	Deep-sea contour current, heavily laden with fine-suspended sediments
Indistinctly layered mud	Poorly sorted mud, exhibiting indistinct, discontinuous layering and discontinuous trains of horizontally oriented coarse grains; thickness variable; bioturbation common; boundaries sharp or gradational	Deep-sea contour current
Biotrubated mud	Poorly sorted mud, intensley disturbed by bioturbation; primary structure absent except for diffuse banding; facies thickness variable boundaries poorly defined and irregular	Hemipelagic sedimentation; contour current
Deformed mud	Mechanically deformed mud showing shear-lineation, crenulated laminae, wispy laminae, microfault and swirl structure	Slumping/sliding

Many detailed records of glaciomarine environments have been provided by the geological record. Five examples from tectonically contrasting settings are summarized below.

Continental facies at a convergent plate margin

The Miocene-Pleistocene Yakataga Formation provides exposures that are almost complete for hundreds of metres vertically and laterally along the uplifted rim of the Gulf of Alaska. The Middleton Island sequence provides an excellent illustration of the facies that are preserved when extremely dynamic temperature glaciers, fed by heavy snow in mountains that are being rapidly uplifted advance across the continental shelf, and subsequently recede rapidly (Eyles & Lagoe 1989, 1990). Typical facies are illustrated in Figure 8.19.

Glaciomarine deltaic complex at the margin of a subsiding rift basin

The 702 m long core obtained from the drill-site CIROS-1 in western McMurdo Sound, Antarctica, has provided a unique record of glaciation extending back to earliest Oligocene time

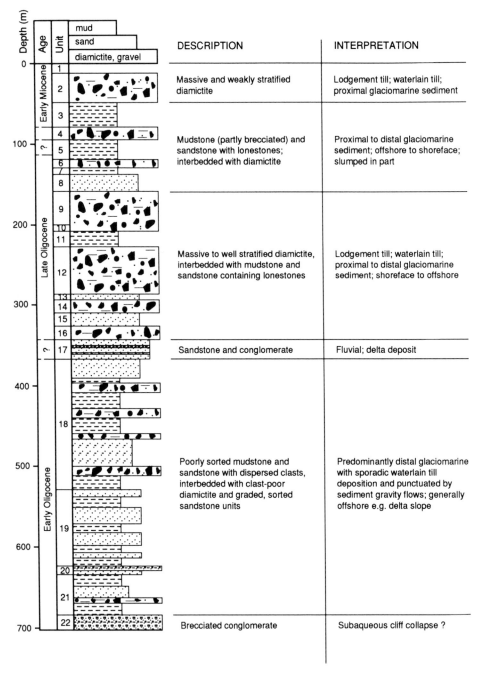

Figure 8.21 Facies association from a subsiding glaciomarine deltaic complex, the CIROS–1 drill-hole, McMurdo Sound, Antarctica. Note that the sequence comprises two distinct parts: the upper one above about 370 m which illustrates periods of ice grounding and lodgement till deposition on the continental shelf, and the lower part which is deeper water and mainly distal glaciomarine. A thin fluvial sequence between is the only indication of subaerial conditions in this core (from Hambrey et al. 1992, with permission of Gebrüde Borntraeger, Stuttgart).

(36 Ma). The dominant facies are diamictite, sandstone and mudstone with abundant flora and fauna (Table 8.2). The facies reflect deposition from temperate glaciers as the Transantarctic Mountains were being uplifted. An early phase of relatively deep-water glaciomarine sedimentation ended with emergence (at about 370 mbsf), and was succeeded by a series of advances and recessions of grounded ice across the continental shelf (Fig. 8.21) (Barrett 1989b).

Glaciomarine facies associated in a subsiding ensialic basin

Thick sequences of interbedded glacioterrestrial and glaciomarine sediments are characteristic of subsiding basins within regions of continental crust. The basins may be linear features, bounded by faults, with connections to the open sea. Much of the North Sea Pleistocene sequence, which reaches a maximum thickness of *c.* 920 m, occupies a linear trough extending north-northwest from the Dutch coast for several hundred kilometres.

The Neoproterozoic glacial era has provided several superbly exposed sequences that were deposited in an ensialic basin, for example, the East Greenland/northeast Svalbard Basin (Fairchild & Hambrey, in press), and the Port Askaig Tillite in Scotland (Spencer 1971, 1985; Eyles 1988b). Glaciomarine sediments are an important component in these mixed terrestrial–marine sequences.

Subsiding passive continental margin dominated by marine ice sheet

Five sites, drilled by Leg 119 of the Ocean Drilling Program on a 180 km long transect across the continental shelf of Prydz Bay to the continental slope, prove an insight into the development of a continental margin under the prolonged influence (over some 40 Ma) of a major ice sheet that flows into the sea (Barron et al. 1989, Hambrey et al. 1991). The principal facies are listed and intepreted in Table 8.2. Seismic records, together with borehole information indicate that there is (a) an upper, flat-lying sequence of grounded-ice and floating-ice sediments, with periods of compaction by ice-sheet loading indicated by overconsolidated sediments, and (b) a lower prograding sequence, mainly composed of glaciomarine sediments.

Ocean basin influenced by iceberg-rafting

Most of our knowledge concerning ice-influenced sedimentation in the deep ocean basins comes from the Ocean Drilling Program (ODP) and its forerunner the Deep Sea Drilling Project (DSDP), operated by the USA on behalf of the international scientific community. Many sites have been drilled in the polar oceans, and a wide range of facies have been documented. The sediments generally comprise a mixture of pelagic and terrigenous components, although the relative proportions are extremely variable. On the one hand the bulk of the sediment may consist of diatom ooze (in the coldest waters) or calcareous ooze (in warmer conditions), with a small ice-rafted component that reflects the geology of the onshore glacierized terrain. On the other hand, microfossils may form a minor component, with terrigenous mud, brought in by bottom currents or as turbidites, being dominant. Again, larger fragments represent ice-rafted material. Two examples are given here.

In the first case, ODP Site 745 lies at the base of the southern slope of the Kerguelen Plateau in 4082 m of water (Barron et al.1989, Ehrmann 1991). The site today is under the influence of Antarctic Bottom Water, derived from beneath the Antarctic ice shelves. Throughout the 215 m thick section, which spans the uppermost Miocene to Quaternary interval, both terrigenous and marine components are present, principally diatom ooze, and silt and clay. Much of the core shows a clear alternation of clayey diatom ooze and diatomaceous clay (sometimes with minor silt) on a scale of decimetres to metres. Pebbles and granules are scattered throughout, but there are few sedimentary structures. These alternations may reflect a greater supply of terrigenous material when the ice sheet was closer to the shelf edge, that is, reflecting cyclicity in the behaviour of the ice sheet.

ODP Site 645 lies low down on the continental slope off southern Baffin Island in a water depth of 2020 m. Here, the sea floor was penetrated by coring to a depth of 1147 m and an early Miocene to Quaternary record was recovered. The influence of ice is perceived above 753 m, in late Miocene sediments, in the form of quartz grains and small pebbles (dropstones). The predominant facies are muddy sand, sandy mud and detrital-carbonate mud (with 30–40% carbonate). As in Antarctica, these facies form alternations on the decimetre to metre scale (Fig. 8.22), a cyclicity supposed to represent glacial periods with heavy ice-rafting, and interglacials without.

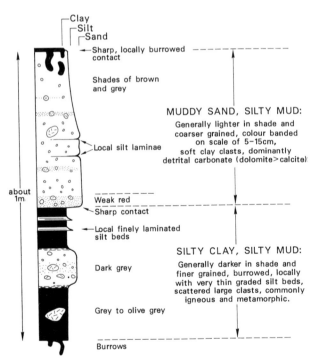

Figure 8.22 Composite sketch of the general features of a "typical" lithological cycle from ODP site 645, Baffin Bay. These cycles are believed to represent interglacial/glacial phases, the latter being indicated by the scattered clasts of ice-rafted material. (Note that the use of "mud" here differs from more normal usage.) (From Srivastava et al. 1987, with permission of Ocean Drilling Program, Texas.)

8.10 Stratigraphic architecture and depositional models of glacially influenced continental shelves

In this section, examples of continental shelf stratigraphic architecture from the Antarctic and Arctic are given (Fig. 8.23). Several other examples from middle latitudes have also been published, for example, off northwest Scotland (Stoker 1990), southeast Canada (King & Fader 1987) and the Gulf of Alaska (Carlson et al. 1990).

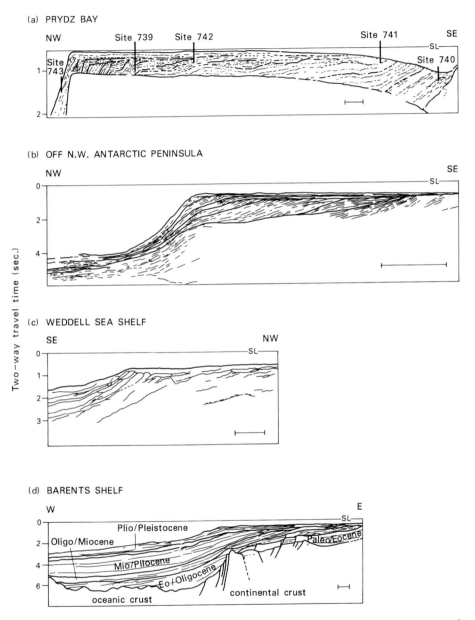

Figure 8.23 Stratigraphic architecture of polar continental shelves. Note the common occurrence of prograding wedges (of which at least the upper parts are probably glacial), overlain by ice-dominated flat-lying sequences (summary diagram, based on various sources, is from Hambrey et al. 1992, with permission of Gebrüder Borntraeger, Stuttgart).

8.10.1 Antarctica

Several features distinguish the Antarctic from other continental margins:

(a) the Antarctic continental shelf has a water depth greater than other shelves, mainly because of ice-sheet loading, with the result that bottom sediments remain unaffected by wave action;

(b) shelf profiles indicate shallowing towards the shelf break as a result of greater glacial erosion of inshore areas, and accretion of glacial sediment on the outer shelf, in addition to loading;

(c) the shelf comprises a lower prograding sequence of normal compaction – overlying this is a flat-lying sequence comprising over-compacted sediments with high seismic velocities;

(d) the shelf today is relatively starved of sediment.

The Prydz Bay shelf is probably the best known high-latitude continental margin as the architecture of the shelf has been determined from a combination of drilling and seismic surveys (Barron et al. 1989, Stagg 1985, Cooper et al. 1991a,b, Hambrey et al. 1991; Fig. 8.23a). As noted in §8.4, a transect across the shelf shows a continuous but complex prograding sequence, covered by a flat-lying sequence. In the inner 120 km of the shelf, the prograding sequence is dominated by fluviatile-deltaic sediments of Mesozoic to early Cretaceous age (Turner 1991). The outer 70 km of the shelf is dominated by a massive diamictite apron, which attains a thickness of at least 400 m, and is interpreted as waterlain till deposited close to the grounding-line as the extended ice sheet decoupled from its bed at the palaeo-shelf break. The glacigenic part of the prograding sequence comprises discrete sedimentary packages, some of which are truncated at the top, whereas others pass landwards into parts of the flat overlying sequence. The drilled part of the shelf shows two distinct glacigenic sequences: a lower gently dipping one with strong syn-sedimentary or possibly glaciotectonic deformation at the base, and an upper steeper prograding sequence with many signs of slumping. Some sedimentary packages (not drilled) show signs of large-scale continental margin collapse and slumping. The topmost parts of the prograding sequence are glaciotectonically deformed and overconsolidated.

The flat-lying sequence, which overlies the prograding sequence, consists mainly of massive diamictite, interpreted as largely the result of deposition from grounded ice. Some of it is overconsolidated, indicating that it was affected by loading, probably during a succession of ice advances. The flat-lying sequence generally thickens seawards from a few metres to around 250 m in the outer parts of the shelf, and illustrates successive stacking of separate units towards the shelf break. Each unit represents a distinct depositional phase. The flat-lying sequence is thickest beneath the banks of the outer shelf, and thinnest in a major channel that extends transversally across the shelf. Some of these separate units pass laterally seawards into prograding packages, including the topmost one.

Seismic data from other parts of Antarctica indicate broadly similar features to those in Prydz Bay. The continental shelf of the Weddell Sea is underlain by a prograding sedimentary sequence, the outer part of which is inferred to be glacigenic, truncated by repeated grounding of the Filchner Ice Shelf (Fig. 8.23c). On top of it is a flat-lying sequence (Hinz &

266

Kristoffersen 1987) also of mainly glacial origin. The same holds for the Pacific margin of the Antarctic Peninsula (Larter & Barker 1989; Fig. 8.23b). In both cases the uppermost sediments are over-compacted, and several ice-advance and recessional cycles are evident. The margin of the Antarctic Peninsula differs from Prydz Bay in being significantly steeper (14° compared with 4°), yet displays no evidence of slumping or collapse.

Model for the Cenozoic development of the Antarctic continental shelf
Based on the stratigraphic architecture of Prydz Bay, which is probably typical of many parts of the continental margin of Antarctica, a generalized model from the time of onset of glacierization at sea level has been developed (Hambrey et al. 1992).

Prior to the onset of glacierization at sea level, the Antarctic continental shelf prograded under the influence of fluvial and deltaic processes (Fig. 8.24a). As the ice sheet developed, it depressed the land isostatically, and the shelf became flooded before the ice advanced across it. The thicker ice over the inner part of the shelf may have led to a reversal of the shelf gradient and, as grounded ice extended to the edge of the shelf, loading increased.

The grounded ice reached, and became decoupled at, the palaeoshelf break by early Oligocene time, and began to deposit waterlain till on the upper parts of the continental slope (Fig. 8.24b). As the ice flowed over its bed it deformed the underlying sediment, and deposited a layer of till. The till itself may have acted as a deforming and erosive medium, and because of the shearing motion induced by the moving ice, some of it may have been displaced towards the grounding-line, thereby contributing to deposition on the palaeoslope. To this was added the material melting out of floating ice, creating a diamicton apron. Slumping was common on this prograding diamict apron.

During the long period characterized by advances to the shelf break with shelf progradation, and recessions with sediment starvation, the morphology of the shelf was modified. Erosion of the inner shelf (with its soft sediment) took place during glacial maxima, whereas till accreted further out as the ice advanced to the shelf break. The till accreted progressively as a flat-lying unit, building upwards and outwards towards the shelf break. The combination of these processes in Prydz Bay is reflected in the tilt of the shelf from an average of about 400 m below sea level at the shelf break to over 800 m on the shelf immediately off the coast.

At the times of full glacial conditions, the only constraint on the advance of ice was the shelf break. Once ice reached this limit, the subsequent advance of grounded ice was probably controlled by deposition and progradation at the grounding-line. It is not thought likely that ice shelves bordering the open sea were as extensive as those of today, although deposition from floating ice is evident from the sedimentary record (Fig. 8.24c).

During phases of ice-recession, such as the present day, most of the Antarctic continental shelf was influenced only by icebergs and sea ice (Fig. 8.24d). Today may be typical of other "interglacials" or periods of reduced ice cover during the Quaternary period. Under such conditions, ice floated in the form of an ice shelf over the inner continental shelves, especially where constrained within embayments such as Prydz Bay, the Ross Sea and the Weddell Sea. Most sediment was released at the grounding-line, where strong bottom-melting probably occurred, as under the Amery Ice Shelf (Robin 1983). However, any sediment not released at the grounding-line would have been protected from melt-out by basal freeze-on, an impor-

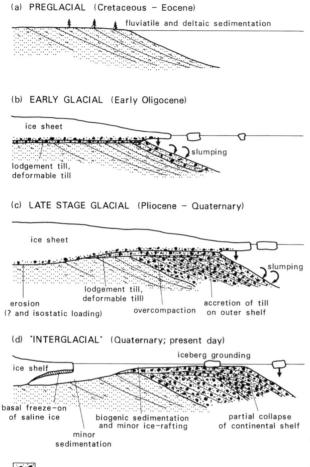

(a) PREGLACIAL (Cretaceous – Eocene)

fluviatile and deltaic sedimentation

(b) EARLY GLACIAL (Early Oligocene)

ice sheet

slumping

lodgement till,
deformable till

(c) LATE STAGE GLACIAL (Pliocene – Quaternary)

ice sheet

slumping

lodgement till,
deformable till)
erosion
(? and isostatic loading)   overcompaction   accretion of till
on outer shelf

(d) "INTERGLACIAL" (Quaternary; present day)

iceberg grounding

ice shelf

basal freeze–on
of saline ice   biogenic sedimentation   partial collapse
and minor ice–rafting   of continental shelf
minor
sedimentation

glaciogenic sediment (mainly diamictite)

preglacial (fluviatile) sediment (mainly sandstone and mudstone)

Figure 8.24 Model for the development of the Antarctic continental shelf, based on drill-hole and seismic data, as exemplified by Prydz Bay. This sequence of events may be inferred for continental margins generally that have been under the prolonged influence of ice (from Hambrey et al. 1992, with permission of Gebrüder Borntraeger, Stuttgart).

tant process, known to occur under the Amery Ice Shelf (Morgan 1972, Budd et al. 1982) and the Filchner–Ronne Ice Shelf (Lange & MacAyeal 1986) today. Such debris thus remained in the ice until the icebergs, which calved from the ice shelf, disintegrated, possibly hundreds of kilometres from their source. Sedimentation on the continental shelves in such circumstances would have been strongly, and perhaps exclusively, diatomaceous, as in many parts of Antarctica today.

In contrast to Pleistocene conditions, the earlier glacial record (Oligocene to Miocene and possibly Pliocene) was characterized by warmer (probably temperate) ice sheets, and the interglacials may have been cool-temperate with vegetation (e.g. the beech *Nothofagus*). Marine sediments may even have been deposited in a seaway across the middle of the continent (Webb et al. 1984). The preservation of interglacial muds on the shelf has not been as effective as that of diamictite, however.

8.10.2 Arctic

A Northern Hemisphere analogue of Prydz Bay is the southwest margin of the Barents Shelf (Vorren et al. 1988, 1990), but its glacial record is much shorter, with regional glaciation starting only about 0.8 Ma (although older glaciomarine sediments may exist). According to Solheim & Kristoffersen (1984), the Barents Shelf was covered by ice sheets five to ten times during Late Cenozoic time. The data come from a combination of seismic data and shallow cores and boreholes through sea-bottom sediments. Broadly, the architecture of the shelf is characterized by a prograding wedge partly capped by a flat-lying glacial sequence. These sediments are underlain in the south by faulted and tilted sedimentary rocks, mainly of Mesozoic age, while at the continental slope a wedge of supposedly Cenozoic age occurs. The shelf area is dominated by a broad submarine valley (probably glacially over-deepened), the Bjørnøyrenna (Bear Island Trough), which attains a depth of 500 m, and which drained much of the ice that covered the Barents Shelf. At the edge of the shelf is a huge trough-mouth fan, similar in size to that in Prydz Bay and on which gravity sliding was an important process. A seismically interpreted profile across the continental shelf and trough-mouth fan is shown in Figure 8.23e (Vorren et al. 1990). Unit *TeE* was regarded as the glacigenic part of the sequence, and the regional unconformity at its base was ascribed to glacial erosion, and considered to represent the glacial/preglacial boundary. The age of this boundary was tentatively put at 0.8 Ma when a change to larger ice volumes became apparent (Vorren 1990). Unit *Te* comprises four major subunits showing stepwise progradation of the palaeoshelf edge. Progradation of the continental shelf probably took place during glacial periods when the Barents Ice Sheet extended close to the shelf edge. Intervening interglacials were probably characterized by sediment starvation at and beyond the shelf edge. Assuming 0.8 Ma is the age of the base of Unit *TeE* shelf, the rate of progradation was about 30 kilometres per million years, i.e. approximately 10 times faster than in preglacial times (Vorren et al. 1990).

Sediments from a borehole through the superficial layer on the shelf comprise overconsolidated glacigenic diamicts. Other facies may be related to proximal prograding diamict-apron deposits, or derivation from meltwater streams associated with ice fed from the Scandinavian highlands to the south.

The geometry of the outer shelf units provides few clues as to the nature of the sediments. However, the suggestion that ice reached the palaeoshelf break on several occasions implies that the slope sediments also belong to a diamict-apron complex, a view supported by the Prydz Bay data.

8.11 Preservation potential of glaciomarine sediments

Deposition on continental shelves provides the best opportunity for glacigenic sediments to be preserved in the geological record. Hence, most sequences from pre-Pleistocene glacial periods tend to be influenced, if not dominated, by continental shelf deposits. Shelf sequences generally show hiatuses, because of repeated advances of ice across the shelf. Erosion is most intense on the inner shelf where the ice is thicker and the early part of the glacial record may have been lost. The outer shelf also loses part of its record as ice advances across it, but considerable build-up of diamictite may occur there from grounded ice and as glaciomarine sediment may occur there. Such areas provide a long but discontinuous record of sedimentation. The most complete direct record of glaciation occurs within the prograding sequences that underlie the outer shelf. These accumulate preferentially during periods when ice is grounded to the continental shelf break. Interglacial periods may be represented by sediment starvation on the palaeoshelf slope, and subject to current erosion, so that the sedimentary record shows a bias towards glacial conditions.

The tectonic setting of many high-latitude continental shelves facilitates the preservation of glaciomarine sediments. Many are slowly subsiding and they allow hundreds, perhaps thousands, of metres of glacigenic sediment to accumulate, e.g. the Ross Sea, Weddell Sea and Prydz Bay in Antarctica. A rather different picture emerges from the Gulf of Alaska. Here, several kilometres of sediment have also accumulated, but primarily because of the huge supply from the rapidly uplifting mountains bordering the Gulf. Continental shelf sequences have been thrust up to form spectacularly well exposed outcrops along the coast and on offshore islands. In contrast to continental shelf areas, deep-ocean sediments may preserve a continuous history of glaciation, but since such sediments lack a direct glacial component other than ice-rafted debris, the record may be difficult to interpret. Furthermore, since oceanic crust and associated sediments older than 200 Ma have only been recorded in fragments, most having been subducted, the longer term preservation of deep-sea glacigenic sediments is poor.

Glossary

ablation The process of wastage of **snow** or ice, especially by melting.

ablation area/zone That part of the glacier surface, usually at lower elevations, over which ablation exceeds accumulation.

abrasion The wearing down of rock surfaces by rubbing and the impact of debris-rich ice.

accumulation area That part of the glacier surface, usually at higher elevations, on which there is net accumulation of **snow**, which subsequently turns into **firn** and then **glacier ice**.

Antarctic Bottom Water A cold sea-bottom current, originating beneath ice shelves in Antarctica, and flowing northwards at depth to influence the world's oceans.

areal scouring Large-scale erosion of bedrock in lowland areas by **ice sheets**.

arête (from the French) A sharp, narrow, often pinnacled ridge, formed as a result of glacial erosion from both sides.

aufeis (from the German) River ice that forms as a result of continued discharge of water from a **glacier** after the winter freeze-up has begun. It comprises continuous sheets of columnar crystals of ice.

avulsion The process of channel-switching in a braided-river system.

axial planar relationship Referring to parallelism with the axial plane of a fold structure.

basal glide Deformation of an ice crystal along discrete bands called **basal planes**.

basal plane The plane within the hexagonal ice crystal that is normal to the optic axis.

basal shear stress The force exerted by an ice mass on its bed.

basal sliding The sliding of a **glacier** over bedrock, a process usually facilitated by the lubricating effect of meltwater.

basket-of-eggs topography Extensive low-lying areas covered by small elongate hills called **drumlins**.

bedrock flutes Ridges, rounded in cross section, formed parallel to the direction of ice movement.

bergschrund (from the German) An irregular crevasse, usually running across an ice slope in the accumulation area, where active **glacier ice** pulls away from ice that adheres to the steep mountainside.

bergstone mud Homogeneous sandy mud containing scattered gravel, deposited mainly from suspension and from **icebergs** in a glacier-influenced fjord or bay.

bergy bit A piece of floating **glacier ice** up to several metres across, commonly derived as a result of disintegration of **icebergs**.

boudin (from the French for sausage) A sausage-shaped block of less ductile material separated by a short distance from its neighbours within a more ductile medium. Boudins normally form perpendicular to the maximum compressive stress.

boulder bed A bed of glacigenic sediment in which the concentration of boulders is exceptionally high (e.g. 50% or more).

boulder clay An English term for **till**, no longer favoured by glacial geologists.

boulder pavement A concentration of **striated** boulders at the top of a bed of poorly sorted sediment that collectively have been planed off by overriding ice. Pavements are formed as a result of glacial abrasion on previously deposited **till** in a terrestrial environment, or by touch-down of ice on to the sea floor in a glaciomarine setting.

braided outwash fan A fan-shaped feature comprising glaciofluvial sediment, emanating from a terrestrial **glacier**.

braided stream A relatively shallow stream with many branches that commonly recombine and migrate across a valley floor. Braided streams typically form down stream of a **glacier**.

breached watershed A short, glacially eroded valley, linking two major valleys across a mountain divide.

breccia Coarse, angular fragments of broken rock or ice, cemented or frozen together into a solid mass.

brown ice Brown-coloured sea ice, commonly found around Antarctica. The colour is due to finely disseminated diatom fragments.

271

calving The process of detachment of blocks of ice from a **glacier** into water.

cavetto form A channel cut into steep rock faces orientated parallel to the valley sides. They are the product of glacial abrasion and/or meltwater. Overhanging upper lips and **striations** are typical.

cavitation Growth and collapse of bubbles in a fluid (e.g. subglacial meltwater) in response to pressure changes. Bubble collapse generates shock waves which result in enhanced erosion of subglacial channels.

channel bar An elongate ridge of sediment between channels in a braided-river system. It usually shows fining of sediment in a downstream direction.

chattermarks A group of crescent-shaped friction cracks on bedrock, formed by the juddering effect of moving ice.

chute A vertical groove in solid bedrock, formed as a result of meltwater erosion at the ice/bedrock contact.

cirque (from the French) An armchair-shaped hollow with steep sides and backwall, formed as a result of glacial erosion high on a mountainside, and often containing a rock basin with a **tarn** (cf. **corrie**, **cwm**).

cirque glacier A **glacier** occupying a **cirque**.

coarse-tail grading A sedimentary bed in which the size of the largest clasts decreases upwards, while the size of the remaining material remains constant.

col (from the French) A high-level pass formed by glacial breaching of an **arête** or mountain mass.

cold glacier A **glacier** in which the bulk of the ice is below the pressure melting point, although ice at the surface may warm up to the melting point in summer, while ice at the bed may also be warmed as a result of geothermal heating.

cold ice Ice that is below the pressure melting point, and therefore dry.

compressing flow The character of ice-flow where a **glacier** is slowing down and the ice is being compressed and thickened in a longitudinal direction.

coriolis force The inertial force associated with variation in the tangential component of the velocity of a particle. It results in an apparent deflection in the centrifugal force generated by the rotation of the Earth.

corrie (from the Gaelic *coire*) A British term for **cirque**.

cover moraine A patchy, thin layer of **till**, revealing, in part, the bedrock topography.

crag-and-tail A glacially eroded rocky hill with a tail of **till** formed down-glacier of it.

creep Permanent deformation of a material under the influence of stress.

crescentic fracture A crescent-shaped crack resulting from friction between debris-rich ice and bedrock; it commonly occurs in groups aligned parallel to ice-flow direction.

crescentic gouge A crescent-shaped scallop, usually several centimetres across, formed as a result of bedrock fracture under moving ice.

crevasse A deep V-shaped cleft formed in the upper brittle part of a **glacier** as a result of the fracture of ice undergoing extension.

crevasse-filling A **crevasse** formed at the base of a **glacier** and filled with soft sediment.

crevasse traces Long veins of clear ice a few centimetres wide, formed as a result of fracture and recrystallization of ice under tension without separation of the two walls; these structures commonly form parallel to open **crevasses** and extend into them. Thicker veins of clear ice resulting from the freezing of standing water in open crevasses are also called crevasse traces.

cumulative (also **total** or **finite strain**) The total amount of strain that a material (e.g. rock, ice) has undergone, usually in response to the prolonged application of stress.

cupola-hill An isolated hill with no obvious source-depression, but having the general characteristics of an assemblage of ice-thrust masses.

curved (winding) channel Sinuous channel cut into bedrock, usually by a subglacial stream under high pressure.

cwm The Welsh term for **cirque**, also sometimes used more generally outside Wales.

cyclopel A graded silt–mud couplet (laminated) formed by tidal processes operating in a glacier-influenced **fjord**.

cyclopsam A graded sand–mud couplet (laminated) formed by tidal processes operating in a glacier-influenced **fjord**.

debris-flow A type of gravity-flow involving the movement on inhomogeneous, unconsolidated material down a slope as a slurry. Total disaggregation of the material takes place, but little fine material is carried away in suspension.

de Geer moraines A group of **moraines** formed subglacially, transverse to flow, beneath a **glacier** terminating in a lake. They have the form of discrete narrow ridges, which may be well spaced.

delta-fan complex A prograding sequence of glaciomarine sediment, gravity-flow material and basal glacial debris, formed at the **grounding-line** of an **ice shelf** or **ice stream** on the continental shelf.

delta moraine (also **kame delta**) A delta complex formed in an ice-frontal position, commonly below water level, where subglacial streams enters a proglacial lake or the sea. Such features are composed of a mixture of glaciofluvial and **glacial debris**.

diagenesis The changes that occur in a sedimentary sequence following deposition (e.g. the alteration of **snow** to **glacier ice** or soft sediment to rock).

diamict A non-sorted terrigenous sediment containing a wide range of particle sizes. Embraces both **diamictite** (lithified) and **diamicton** (unconsolidated).

diamict apron (cf. **till delta**) A prograding sequence comprising **diamict** and reworked sediments, formed just seawards of the **grounding-line/zone** at a break in slope on the sea floor.

diffluence The processes whereby ice in one valley overflows a **col** or group of **cols** into an adjacent valley.

dilation The reduction in volume of a material. Commonly occurs in bedrock following ice-removal, and leads to the development of cracks.

dirt-cone A thin veneer of debris, draping a cone of ice up to several metres high, formed as a result of the debris locally retarding ablation of the glacier surface.

diurnal variation Variations taking place on a day–night cycle.

dome A smooth, rounded boss of glacially abraded bedrock, commonly exceeding hundreds of metres in diameter.

dropstone A relatively large clast that falls through the water column into soft sediment, disrupting the bedding or laminae. Draping of sediment over the top of the clast subsequently occurs. In a glacial context, dropstones are released from icebergs.

drumlin (from the Gaelic) A streamlined hillock, commonly elongated parallel to the former ice-flow direction, composed of **glacial debris**, and sometimes having a bedrock core; formed beneath an actively flowing **glacier**.

drumlinoid ridges (**drumlinized ground moraine**) Elongate, strongly linear ridges, intermediate between **drumlins** and fluted **moraine**.

englacial debris Debris dispersed throughout the interior of a **glacier**, derived either from the surface through burial in the **accumulation area** and through falling into **crevasses**, or from the uplifting of basal debris by thrusting processes.

englacial stream A meltwater stream that has penetrated below the surface of a **glacier**, and is making its way towards the bed.

erratic A boulder or large block of bedrock that is being or has been transported away from its source by a **glacier**.

esker (from the Gaelic) A long, commonly sinuous ridge of sand and gravel, deposited by a stream in a subglacial tunnel.

equilibrium line/zone The line or zone on a **glacier** surface where a year's ablation balances a year's accumulation (cf. **firn line**). It is determined at the end of the **ablation** season, and commonly occurs at the boundary between **superimposed ice** and **glacier ice**.

exfoliation The process of removal of sheets of bedrock along joints parallel to the rock surface.

extending flow The character of ice-flow where a **glacier** is accelerating and the ice is being stretched and thinned in a longitudinal direction.

facies (singular & plural) A sediment type characterized by an assemblage of features, including lithology, texture, sedimentary structures, fossil content, geometry, bounding relations. **Lithofacies** refers to a particular lithological type.

facies architecture The large-scale two- or three-dimensional geometry of facies associations, usually on a basin-wide scale and revealed in seismic profiles.

facies association The grouping of **facies** into an environmentally coherent assemblage, e.g. a glaciomarine facies association, usually applied to vertical sedimentary logs.

fault A displacement in a **glacier** formed as a result of fracture of the ice without separation of the walls. It is recognized by the discordance of layers in the ice on either side of the fracture. A **normal fault** is a high-angle fracture resulting from the maximum compressive stress acting vertically and the intermediate and least-compressive stresses horizontally. A **thrust fault** is a low-angle fracture in which the maximum compressive stress acted horizontally and the least compressive stress vertically. A **strike–slip fault** is a fracture showing sideways displacement where both maximum and minimum compressive stresses both acted horizontally.

finite strain (see **total strain**).

firn (from the German) Dense, old **snow** in which the crystals are partly joined together, but in which the air pockets still communicate with each other.

firn line The line on a **glacier** that separates bare ice from **snow** at the end of the **ablation** season.

fjord (from the Norwegian; **fiord** in North America and New Zealand) A long, narrow arm of the sea, formed as a result of erosion by a **valley glacier**.

floe (see **mega-block**).

floe (see **raft**).

fluted moraine (flutes) Rounded, strongly linear ridges, up to a few metres in width and height, usually formed in association with **lodgement till** on land. Similar features have been recorded in shallow glaciomarine settings.

foliation Groups of closely spaced, often discontinuous, layers of coarse bubbly, coarse clear and fine-grained ice, formed as a result of shear or compression at depth in a **glacier**.

fold Layers of ice that have been deformed into a curved form by flow at depth in a **glacier**. **Isoclinal folds** have parallel limbs and thickened hinges. **Similar folds** have a thickened hinge and thinned limbs. **Parallel folds** maintain a uniform layer-thickness around the fold. **Recumbent folds** are those with near-horizontal axes and are commonly associated with **thrust** faults.

gendarme (from the French for policeman) A pinnacle of rock on the crest of a narrow ridge, especially an **arête**.

gentle hill A mound of **till** resting on a detached block of bedrock.

geothermal heat The heat output from the Earth's surface. This affects **glaciers**, especially in the polar regions, by warming the basal zone to the **pressure melting point**.

gilbertian delta A delta produced by a stream with a high bedload entering a quiet water body. Such deltas have steep slopes, and avalanching causes the delta to advance and create a single cross-bedded set of sand and gravel.

glacial debris Material in the process of being transported by a **glacier** in contact with **glacier ice**.

glacial drift A general term embracing all rock material deposited by **glacier ice**, and all deposits of predominantly glacial origin deposited in the sea from **icebergs**, and from glacial meltwater.

glacial period/glaciation A period of time when large areas (including present temperate latitudes) were ice-covered. Many glacial periods have occurred within the past few million years, and are separated by **interglacial periods**.

glacial trough A valley or **fjord**, often characterized by steep sides and a flat bottom, with multiple basins, resulting primarily from **abrasion** by strongly channelled ice.

glaciated The character of land that was once covered by **glacier ice** in the past (cf. **glacierized**).

glacier A mass of ice, irrespective of size, derived largely from **snow**, and continuously moving from higher to lower ground, or spreading over the sea.

glacier ice Any ice in, or originating from, a **glacier**, whether on land or floating on the sea as **icebergs**.

glacierized The character of land currently covered by **glacier ice** (cf. **glaciated**).

glacier karst Debris-covered stagnant ice, sometimes found at the **snout** of a retreating **glacier**, with many lake-bearing caverns and tunnels.

glacier sole The lower few metres of a (usually sliding) **glacier** which are rich in debris picked up from the bed.

glacier table A boulder sitting on a pedestal of ice, resulting from the protective effect of the rock mass on **ablation** of the ice surface during sunny weather.

glacier tongue (see **ice tongue**).

glacigenic sediment Sediment of glacial origin. The term is used in a broad sense to embrace sediments with a greater or lesser component derived from **glacier ice**.

glaciomarine sediment A mixture of **glacigenic** and marine sediment, deposited more or less contemporaneously.

glaciotectonic deformation (glaciotectonism) The process whereby subglacial and proglacial sediment and bedrock is disrupted by ice-flow. It is usually manifested in the form of distinct topographic features in which folds and thrusts are commonplace.

Glen's Flow Law The empirical relationship which describes the manner in which ice deforms in response to an applied stress. First proposed by the British physicist John Glen, following experimental studies in the early 1950s.

gravity flowage The process of transport of unconsolidated sediment down a slope, either subaerially or subaquatically. The term embraces **debris-flow** and **turbidity currents** at opposite ends of the spectrum.

groove A glacial abrasional form, with striated sides and base, orientated parallel to the ice-flow direction. Grooves are often parabolic in cross section and up to several metres wide and deep.

grounding-line (or **grounding-zone**) The line or zone at which an ice mass enters the sea or a lake and begins to float, e.g. in the inner parts of an **ice shelf** or an **ice stream**.

grounding-line fan A subaquatic fan made up of material emerging from a subglacial tunnel where a **glacier** terminates in the sea. Comprises a complex range of facies.

hanging glacier A **glacier** that spills out from a high-level **cirque** or clings to a steep mountainside.

hanging valley A tributary valley whose mouth ends abruptly part way up the side of a trunk valley, as a result of the greater amount of glacial down-cutting of the latter.

highland ice field A near-continuous stretch of **glacier ice**, but with an irregular surface that mirrors the underlying bedrock, and punctuated by **nunataks**.

hill–hole pair A combination of an ice-scooped basin and a hill of ice-thrust, often slightly crumpled material of similar size, resulting from glaciotectonic processes.

horn A steep-sided, pyramid-shaped peak, formed as a result of the backward erosion of cirque **glaciers** on three or more sides.

hummocky (ground) moraine Groups of steep-sided hillocks, comprising **glacial drift**, formed by dead-ice-wastage processes. Some hummocky **moraines** may be arranged in a crude transverse-to-valley orientation and may reflect thrusting processes in the **glacier snout**. Both terrestrial and marine types are known.

ice age A period of time when large ice sheets extend from the polar regions into temperate latitudes. The term is sometimes used synonymously with **glacial period**, or embraces several such periods to define a major phase in the Earth's climatic history.

ice apron A steep mass of ice, commonly the source of ice avalanches, that adheres to steep rock near the summits of high peaks.

iceberg A piece of ice of the order of tens of metres or more across that has been shed by a **glacier** into a lake or the sea.

iceberg turbate Sediment disturbed and added to by **icebergs** grounding on a sea or lake floor.

ice cap A dome-shaped mass of **glacier ice**, usually situated in a highland area, and generally defined as covering $< 50\,000\,\text{km}^2$.

ice cliff (ice wall) A vertical face of ice, normally formed where a **glacier** terminates in the sea, or is undercut by streams. These terms are also used more specifically for the face that forms at the seaward margin of an **ice sheet** or **ice cap** and which rests on bedrock at or below sea level.

ice-contact delta Delta formed from a **subglacial stream** where a **glacier** terminates in a standing body of water. The top surface is in the intertidal zone.

icefall A steep, heavily crevassed portion of a **valley glacier**.

ice-marginal channel A water channel formed at the margin of a (usually cold) **valley glacier**, inside the **lateral moraine**, if present.

ice sheet A mass of ice and **snow** of considerable thickness and covering an area of more than $50\,000\,\text{km}^2$.

ice shelf A large slab of ice floating on the sea, but remaining attached to and partly fed by land-based ice.

ice stream Part of an **ice sheet** or **ice cap** in which the ice flows more rapidly, and not necessarily in the same direction as the surrounding ice. The margins are often defined by zones of strongly sheared, crevassed ice.

ice stream boundary ridge A submarine ridge of bedrock or soft sediment defining the former boundary between two **ice streams** as they crossed a continental shelf.

275

ice tongue (or **glacier tongue**) An unconstrained, floating extension of an **ice stream** or **valley glacier**, projecting into the sea.

interflow Flow in which water entering a lake or the sea is of the same density as the main body, resulting in ready mixing of the two.

interglacial (period) A period, similar to or warmer than that of today, during which ice has receded mainly to the polar regions and high mountain areas.

internal deformation That component of glacier flow that is the result of the deformation of **glacier ice** under the influence of accumulated **snow** and **firn**, and gravity.

isotopes Varieties of elements, all with identical chemical, but not precisely equal physical, properties.

jökulhlaup (from the Icelandic) A sudden and often catastrophic outburst of water from a **glacier**, such as when an ice-dammed lake bursts or an internal water pocket escapes.

kame (from the Gaelic) A steep-sided hill of sand and gravel deposited by glacial streams adjacent to a **glacier** margin.

kame delta (see **delta moraine**).

kame field A large area covered by many discrete kames.

kame moraine An end-moraine complex that has been substantially reworked by glaciofluvial processes.

kame plateau An extensive area of ice-contact sediments formed adjacent to a **glacier** that has not yet been dissected.

kame terrace A flat or gently sloping plain, deposited by a stream that flowed towards or along the margin of a **glacier**, but left above the hillside when the ice retreated.

kettle (or **kettle hole**) A self-contained bowl-shaped depression within an area covered by glacial stream deposits, and often containing a pond. A kettle forms as a result of the burial of a mass of **glacier ice** by stream sediment and the subsequent melting of the ice.

kinematic wave The means whereby mass-balance changes are propagated down-glacier. The wave has a constant discharge and it moves faster than the ice velocity. Kinematic waves are visible as bulges on the ice surface; when they reach the **snout**, the **glacier** is able to advance.

knock-and-lochan topography (a Scottish term) Rough, ice-abraded, low-level landscape, comprising small hills of exposed bedrock, and rock basins with small lakes and bogs.

laminite A laminated sediment.

large composite-ridge A ridge, or series of ridges, in front of a **glacier** and comprising large slices of upthrust, contorted sedimentary bedrock and drift. They exceed 100 m in height.

lee-side cone (see **crag-and-tail**).

levée A bank of sediment bordering a stream channel, constructed during a period of bankfull discharge.

linear moraine A composite feature on a continental shelf, comprising **till tongue** sediments overlain by subaquatic ice-recessional features.

lithofacies (see **facies**).

Little Ice Age That period of time that led to expansion of valley and **cirque glaciers** worldwide, with their maximum extents being attained in many temperate regions about AD 1700–1850 or around 1900 in Arctic regions.

lodgement The process whereby basal glacial debris is "plastered" on to the substrate beneath an actively moving glacier.

loess Wind-blown sediment of silt grade, often derived as a result of winnowing of fines from glacial outwash plains.

lonestone An isolated clast of pebble or larger size in a fine-grained matrix.

lunate fracture Moon-shaped crescentic fracture with the horns and steeper face at the down-glacier end.

mass balance (or **mass budget**) A year-by-year measure of the state of health of a **glacier**, reflecting the balance between accumulation and **ablation**. A **glacier** with a positive mass balance in a particular year gained more mass through accumulation than was lost through **ablation**; the reverse is true for negative mass balance.

mega-block (also known as a **raft**) A large piece of glacially transported bedrock, buried in drift. Such blocks are thin in relation to their linear dimensions, which may exceed 100 m.

mixton(ite) a synonym for **diamicton(ite)**, but now largely redundant.

morainal bank A bank formed below water in front of a stable ice front in a **fjord** or lake, as a result of **lodgement**, melting, dumping, push and squeeze processes.

moraine Distinct ridges or mounds of debris laid down directly by a **glacier** or pushed up by it. The material is typical **till**, but fluvial, lake or marine sediments may also be involved. Longitudinal moraines include a **lateral moraine** which forms along the side of a **glacier**; a **medial moraine** occurring on the surface where two streams of ice merge; and a **fluted moraine** which forms a series of ridges beneath the ice, parallel to flow. Transverse moraines include a **terminal moraine** which forms at the farthest limit reached by the ice, a **recessional moraine** which represents a stationary phase during otherwise general retreat, and a set of **annual moraines** representing a series of minor winter re-advances during a general retreat. A **push moraine** is a more complex form, developed especially in front of a **cold glacier** during a period of advance.

moulin A water-worn pothole formed where a surface meltstream exploits a weakness in the ice. Many moulins are cylindrical, several metres across, and extend down to the glacier bed, although often in a series of steps.

moulin kame A mound of debris up to several metres high that accumulated in the bottom of a **moulin**.

nail-head striation An asymmetrical, relatively short scratch mark, blunt at the downstream end and tapering towards the upstream end.

nival flood The flood that accompanies the phase of rapid snowmelt in spring/early summer.

nunatak (from the Inuit) An island of bedrock or mountain projecting above the surface of an **ice sheet** or **highland ice field**.

Nye channel (named after a British physicist) A channel cut into bedrock by subglacial meltwater under high pressure. Usually less than 1 m across; commonly deeper than they are wide.

ogives Arcuate bands or waves, with the apex pointing down-glacier, that develop in an **icefall**. Alternating light and dark bands are called **banded ogives** or **Forbes' bands**; **wave ogives** are a second type. Each pair of bands or one wave and trough represents a year's movement through the **icefall**.

outwash plain A flat spread of debris deposited by meltwater streams emanating from a **glacier** (cf. **sandar**).

overconsolidated A geotechnical term to indicate a more indurated sediment than would be expected from the depth observed below the surface. Overconsolidation may be attributable to loading by an ice mass.

overflow A flow of water entering a lake or the sea, having a density less than the main body of water, and therefore remaining at the surface as a distinct layer.

overflow channel A channel cut through a hill or ridge by meltwater, as a result of ponding back by an ice mass.

permafrost Ground that remains permanently frozen. It may be hundreds of metres thick with only the top few metres thawing out in summer.

piedmont glacier (from the French) A **glacier** that spreads out as a wide lobe on leaving the mountains or a narrow trough.

pitted plain A plain of glaciofluvial sediment with many lake-filled depressions (**kettles**), resulting from the melting of buried blocks of **glacier ice**.

plastically moulded (or p-) forms Smooth rounded forms of various types cut into bedrock by the combined effects of the erosive power of ice and meltwater under high pressure.

plough-mark (a) A groove or furrow caused by the impact and movement of **icebergs** across the sea or lake floor, or along a beach under the influence of tides. (b) A groove formed by a stone at the ice/bedrock interface scoring soft sediment and pushing up a small ridge in front.

polynya An area of ocean with sea ice that remains relatively ice-free in winter, due to the upwelling to the surface of warmer water.

portal The open archway that develops when a meltwater stream emerges at the **snout** of a **glacier**.

pressure-melting Enhanced melting resulting from the effect of ice impinging on bedrock, or an object under stress impinging on ice.

pressure release The process of fracturing of bedrock that takes place as a result of the removal of stress as an overlying body of ice melts.

pure shear (non-rotational strain) Deformation of a substance by extension normal to an applied compressive stress and extension parallel to it. This results in no change to the strained area, nor to rotation of the cumulative strain axes.

raft (see **mega-block**).

randkluft (from the German) The narrow gap that develops between a rock face and steep firn and ice at the head of a **glacier**.

rat-tail A minor ridge, parallel to striations, extending down-glacier from a knob of more resistant rock.

regelation ice Ice formed from meltwater as a result of the lowering of pressure beneath a **glacier**.

rejuvenated (or regenerated) glacier A **glacier** that develops from ice-avalanche debris beneath a rock cliff.

rhythmite A sedimentary unit comprising a repetitive succession of sedimentary types, e.g. mud/sand, irrespective of time and relative thickness. A **varve** is one specific type of rhythmite.

ribbon lake A long, narrow lake resulting from glacial erosion, and commonly containing multiple basins.

riegel (from the German) A rock barrier that extends across a **glaciated** valley, usually comprising harder rock, and often having a smooth slope facing up-valley and a rough slope facing down-valley.

rinnen (from the German) (see **tunnel valley**).

roche moutonnée (from French) A rocky hillock with a gently inclined, smooth slope facing up-valley resulting from glacial abrasion, and a steep, rough slope facing down-valley resulting from glacial plucking.

rock basin A lake- or sea-filled bedrock depression carved out by a **glacier**.

rock flour Bedrock that has been pulverized at the bed of a **glacier** into clay- and silt-sized particles. It is commonly carried in suspension in glacial meltwater streams, which take on a milky appearance as a result.

rockslide An accumulation of rock debris resulting from a catastrophic rockfall. Sometimes during the fall, debris becomes airborne and accumulates as a pulverized mass, far distant from the source.

rogen (ribbed) moraine (named after a Swedish lake) Large-scale, transversely orientated, irregular ridges of complex origin, including thrusting.

röthlisberger channel (named after a Swiss glaciologist) A channel incised upwards into the base of a **glacier** by a subglacial meltstream.

sandar (plural **sandur**; from the Icelandic) Laterally extensive flat plains of sand and gravel with **braided streams** of glacial meltwater flowing across them. They are usually not bounded by valley walls and they commonly form in coastal areas.

sea ice Ice that forms by freezing of the sea (cf. **ice shelf** and **icebergs** which float on the sea).

sedimentary model A pictorial representation of the processes and sedimentary products that contribute to a particular sedimentary environment.

sedimentary stratification The annual layering that forms from the accumulation of **snow**, and is preserved in **firn** and sometimes in **glacier ice**.

seismic stratigraphy The application of seismic profiling techniques to the interpretation of the stratigraphy of inaccessible (subsurface) areas, notably offshore.

sérac (from the French) A tower of unstable ice that forms between **crevasses**, often in **icefalls** or other regions of accelerated glacier flow.

sequence stratigraphy The determination of the stratigraphy of sedimentary sequences by linking unconformities of temporal significance (sequence boundaries) to sea-level changes, thereby aiding regional correlation. A development of seismic stratigraphy, but applicable to well exposed onshore sequences.

shear (or thrust) ridge A ridge or line of debris cropping out at the surface of a **glacier**, thrust up from a subglacial position, especially near its **snout**.

shelf moraine A prominent ridge interpreted as an end-moraine complex, formed at the edge or in the middle of a continental shelf.

shelf valley (see **submarine trough**).

Sichelwanne A crescent-shaped depression or scallop-like feature cut into bedrock, probably largely the result of meltwater erosion. The axis is coincident with the ice-flow direction, and the horns of the feature point forwards.

sill A submarine barrier of rock or **moraine** that occurs at the mouth of, or between, rock basins in a **fjord**.

simple shear Deformation of a substance by displacement along discrete (often closely spaced) surfaces or shear planes. The orientations of the principal axes of cumulative strain rotate as deformation proceeds.

This type of deformation can be demonstrated by gradually smearing out a stack of cards.

slope valleys Groups of gullies forming a dendritic pattern on the continental slope associated with subglacial discharge at times when the ice advanced to the continental shelf edge.

slumping The process of downslope mass movement whereby the internal organization of the mass of debris is not totally disaggregated. Deformation may be evident in the form of recumbent fold structures.

small composite-ridge Smaller scale versions (< 100 m in height) of **large composite-ridges** and comprising mainly unconsolidated sediment thrust up into a series of ridges by a **glacier**.

snout The lower part of the **ablation area** of a **valley glacier**.

snow An agglomeration of precipitated ice crystals, most of which are star-shaped and delicate, forming a low-density mass with a high air content.

snow swamp An area of saturated **snow** lying on **glacier ice**. If movement is triggered on a slope, a slush avalanche may develop. The snow swamp may not be visible until one steps into it.

sole (see **glacier sole**).

sole thrust The lowest thrust in a deformed sedimentary sequence that defines the plane below which no further displacements occur.

strain The amount by which an object becomes deformed under the influence of **stress**.

strain-softening The effect whereby, under a constant **stress**, the strain rate increases with time.

stauchmorne (from the German) A collective term for ice-push ridges of the **large** and **small composite-ridge** types.

stress The force applied to an object.

striae Linear, fine scratches formed by the abrasive effect of debris-rich ice sliding over bedrock. Intersecting sets of striae are formed as stones are rotated or if the direction of ice-flow over bedrock changes.

striated The scratched state of bedrock or stone surfaces after ice has moved over them.

subaquatic The state of being under water; in this context commonly refering to glaciolacustrine and glaciomarine environments.

subglacial debris Debris that has been released from ice at the base of a **glacier**. It usually shows signs of rounding due to **abrasion** at the contact between ice and bedrock.

subglacial delta (cf. **till delta, diamict apron**) Prograding wedge of glacial, marine and recycled debris formed seawards of the **grounding-line** of **ice shelves** or **ice streams**.

subglacial gorge A steep, often vertically sided gorge cut into bedrock by a **subglacial stream** under high pressure.

subglacial stream A stream that flows beneath a **glacier**, and which usually cuts into the ice above to form a tunnel.

sublimation The process whereby a material (e.g. ice) passes from the solid to the vapour state directly, without melting intervening.

superimposed drainage A drainage system unrelated to current bedrock structure, inherited from an earlier phase of development on once-overlying bedrock, now removed.

superimposed ice Ice that forms as a result of the freezing of water-saturated **snow**. It commonly forms at the surface of a **glacier** between the **equilibrium line** and the **firn line**, and provides additional mass to the **glacier**.

supraglacial debris Debris that is carried on the surface of a **glacier**. Normally this is derived from rockfalls and it tends to be angular.

supraglacial stream A stream that flows over the surface of a **glacier**. Most supraglacial streams descend via **moulins** into the depths or to the base of a **glacier**.

surge A short-lived phase of accelerated **glacier** flow during which the surface becomes broken up into a maze of **crevasses**. Surges are often periodic and are separated by longer periods of relative inactivity or even stagnation.

tabular iceberg A flat-topped iceberg that has become detached from an **ice shelf**, **ice tongue** or floating **tidewater glacier**.

tabular moraine Irregular morainic complexes formed in an intermediate continental shelf position.

tarn A small lake occupying a hollow eroded out by ice or dammed by a **moraine**; especially common in **cirques**.

temperate glacier (see **warm glacier**).

thermal regime The state of a **glacier** as determined by its temperature distribution.

thrust A low-angle fault, usually formed where the ice is under compression. Thrusts commonly extend from the bed and are associated with debris and overturned folds.

thrust ridge (see **shear ridge**).

tidewater glacier A **glacier** that terminates in the sea, usually in a bay or **fjord**.

till A mixture of mud-, sand- and gravel-sized material deposited directly from **glacier ice**.

till delta A prograding wedge of till that has been transported in conveyor-belt fashion to the **grounding-line** of an **ice shelf** or **ice stream** (cf. **diamict apron**).

tillite The lithified equivalent of **till**.

tilloid A **till**-like deposit, but of uncertain origin (a term rarely used today).

till pellet A lump of **till** that was frozen when incorporated into the host sediment.

till plain A nearly flat or slightly rolling and gently inclined surface, underlain by a near-continuous cover of thick **till**.

till tongue An inclined wedge of **till** and other material associated with the **grounding-line** of an oscillating marine ice margin.

tongue The part of a **valley glacier** that extends below the **firn line**.

total strain (see **cumulative strain**).

transfluence The large-scale breaching of a mountain range by glaciers emanating from an **ice sheet**, with all **cols** being occupied by discharging ice.

transverse ridge (see **de Geer moraines**).

trim line A sharp line on a hillside marking the boundary between well vegetated terrain that has remained ice-free for a considerable time, and poorly vegetated terrain that until relatively recently lay under **glacier ice**. In many areas the most prominent trim lines date from the **Little Ice Age**.

trough-head (or **trough-end**) The steep, plucked transverse rockface at the head of a **glacial trough**.

trough-mouth fan A large-scale prograding, arcuate fan at the mouth of a continental shelf trough. The fan extends out from the shelf edge into the deep-ocean basin.

tunnel valley (Rinnen) A large subglacial, steep-sided channel cut into soft sediment or bedrock by meltwater. The channel may have a reverse gradient in places.

turbidity current A high-density, debris-laden current that flows down a slope. The resulting turbidite is a sediment characterized by graded bedding.

unconformity A discontinuity in the annual layering in **firn** or ice, resulting from a period when **ablation** cut across successive layers.

underflow A flow of water of greater density than the lake or sea into which it flows, thereby descending a slope as a bottom-hugging current or turbidity-flow.

valley glacier A **glacier** bounded by the walls of a valley, and descending from high mountains, an **ice cap** on a plateau, or an **ice sheet**.

valley train A braided-river system that extends across the whole width of a valley between steep-sided mountains.

varve A sedimentary bed or lamina, or sequence of laminae, deposited in a body of still water (usually a lake) and representing one year's accumulation.

varvite A lithified **varve**.

warm (temperate) glacier A **glacier** whose temperature is at the **pressure melting point** throughout, except for a cold wave of limited penetration that occurs in winter.

warm ice Ice that is at the melting point regardless of pressure. The temperature may be slightly below 0°C at the base of a **glacier** where the ice is under high pressure.

whaleback A smooth, scratched, glacially eroded bedrock knoll several metres high, and resembling a whale in profile.

Bibliography

Aber, J. S., D. G. Croot, M. M. Fenton 1989. *Glaciotectonic landforms and structures*. Dordrecht: Kluwer Academic Publishers.

Allen, C. R., W. B. Kamb, M. F. Meier, R. P. Sharp 1960. Structure of the lower Blue Glacier, Washington. *Journal of Geology* **68**, 601–25.

Alley, D. W. & R. M. Slatt 1976. Drift prospecting and glacial geology in the Sheffield Lake – Indian Pond area, northcentral Newfoundland. See Leggett (1976), 249–66.

Alley, R. B., D. D. Blankenship, S. T. Rooney, C. R. Bentley 1989. Sedimentation beneath ice shelves – the view from Ice Stream B. *Marine Geology* **85**, 101–20.

Allison, I. F. 1979. The mass budget of the Lambert Glacier drainage basin, Antarctica. *Journal of Glaciology* **22**, 223–35.

Anderson, J. B. 1983. Ancient glacial-marine deposits: their spatial and temporal distribution. See Molnia (1983b), 10–92.

Anderson, J. B. & G. M. Ashley 1991. *Glacial marine sedimentation; paleoclimatic significance*. Geological Society of America, Special Paper 261.

Anderson, J. B. & L. R. Bartek 1992. Cenozoic glacial history of the Ross Sea revealed by intermediate resolution seismic reflection data combined with drill site information. In *The Antarctic paleoenvironment: a perspective on global change*, 231–63. American Geophysical Union, Antarctic Research Series, vol. 56.

Anderson, J. B. & B. F. Molnia 1989. *Glacial-marine sedimentation*. Washington: American Geophysical Union, Short Course in Geology 9 (28th Int. Geol. Congress).

Anderson, J. B., D. D. Kurtz, E. W. Domack, K. M. Balshaw 1980. Glacial and glacial marine sediments of the Antarctic continental shelf. *Journal of Geology* **88**, 399–414.

Anderson, J. B., C. Brake, E. W. Domack, N. Myers, R. Wright 1983. Development of a polar glacial-marine sedimentation model from Antarctic Quaternary deposits and glaciological information. See Molnia (1983b), 233–64.

Andrews, J. T. & C. L. Matsch 1983. *Glacial marine sediments and sedimentation: an annotated bibliography*. Norwich: GeoAbstracts 11.

Armstrong, T., B. Roberts, C. Swithinbank 1973. *Illustrated glossary of snow and ice*. Cambridge: Scott Polar Research Institute.

Ashley, G. M. 1975. Rhythmic sedimentation in glacial lake Hitchcock, Massachusetts–Connecticut. See Jopling & McDonald (1975), 304–20.

Ashley, G. M. 1989. Classification of glaciolacustrine sediments. See Goldthwait & Matsch (1989), 243–60.

Banerjee, I & B. C. McDonald 1975. Nature of esker sedimentation. See Jopling & McDonald (1975).

Barrett, P. J. 1975. Characteristics of pebbles from Cenozoic marine glacial sediments in the Ross Sea (DSDP Sites 270–274) and the south Indian Ocean. In *Initial Reports, Deep Sea Drilling Project, Leg 28*, D. E. Hayes, L. A. Frakes & Shipboard Scientific Party (eds), 769–84. Washington, DC: US Government Printing Office.

Barrett, P. J. 1980. The shape of rock particles' a critical review. *Sedimentology* **27**, 291–303.

Barrett, P. J. (ed.) 1986. *Antarctic Cenozoic history from the MSSTS-1 drillhole, McMurdo Sound*. Wellington: DSIR, Bulletin 237.

Barrett, P. J. 1989a. Sediment texture. See Barrett (1989a), 49–58.

Barrett, P. J. (ed.) 1989b. *Antarctic Cenozoic history from the CIROS-1 drillhole, McMurdo Sound, Antarctica*. Wellington: DSIR, Bulletin 245.

Barrett, P. J. & M. J. Hambrey 1992. Plio-Pleistocene sedimentation in Ferrar Fiord, Antarctica. *Sedimentology* **39**, 109–23.

Barrett, P. J., A. R. Pyne, B. L. Ward 1983. Modern sedimentation in McMurdo Sound, Antarctica. In *Antarctic earth science*, R. L. Oliver, P. R. James & J. B. Jago (eds), 550–55. Canberra: Australian Academy of Science and Cambridge: Cambridge University Press.

Barrett, P. J., M. J. Hambrey, D. M. Harwood, A. R. Pyne, P-N. Webb 1989. Synthesis. See Barrett (1989b), 241–51.

Barron, J., B. Larsen, Shipboard Scientific Party 1989. *Leg 119, Kerguelen Plateau and Prydz Bay, Antarctica. Proceedings of the Ocean Drilling Program*, vol. 119, part A.

Barron, J., B. Larsen, Shipboard Scientific Party 1991. *Leg 119, Kerguelen Plateau and Prydz Bay, Antarctica. Proceedings of the Ocean Drilling Program*, vol. 119, part B, Scientific Results.

Beuf, S., B. Biju-Duval, O. de Charpal, P. Rognon, O. Gariel, A. Bennacef 1971. *Les grès du Paléozoïque inférieur au Sahara. Sédimentation et discontinuité. Evolution structurale d'un craton*. Paris: Edit. Technip.

Bjørlykke, K., B. Bue, A. Elverhøi 1978. Quaternary sediments in the western part of the Barents Sea and their relation to the underlying Mesozoic bedrock. *Sedimentology* **25**, 227–46.

Boothroyd, J. C. & G. M. Ashley 1975. Processes, bar morphology and sedimentary structures on braided outwash fans, northeastern Gulf of Alaska. See Jopling & McDonald (1975), 193–222.

Borns, H. W. & C. L. Matsch 1989. A provisional genetic classification of glaciomarine environments, processes and sediments. See Goldthwait & Matsch (1989), 261–6.

Bouchard, M. A. 1989. Subglacial landforms and deposits in central and northern Québec, Canada, with emphasis on Rogen moraines. *Sedimentary Geology* **62**, 293–308.

Boulton, G. S. 1967. The development of a complex supraglacial moraine at the margin of Sørbreen, Ny Friesland, Vestspitsbergen. *Journal of Glaciology* **6**, 717–36.

Boulton, G. S. 1968. Flow tills and related deposits on some Vestspitsbergen glaciers. *Journal of Glaciology* **7**, 391–412.

Boulton, G. S. 1970. The deposition of subglacial and melt-out tills at the margins of certain Svalbard glaciers. *Journal of Glaciology* **9**, 231–45.

Boulton, G. S. 1971. Till genesis and fabric in Spitzbergen, Svalbard. In *Till, a symposium*, R. P. Goldthwait (ed.), 41–72. Columbus: Ohio State University Press.

Boulton, G. S. 1972. Modern Arctic glaciers as depositional models for former ice sheets. *Quarterly Journal of the Geological Society of London* **128**, 361–93.

Boulton, G. S. 1978. Boulder shapes and grain-size distribution of debris as indicators of transport paths through a glacier and till genesis. *Sedimentology* **25**, 773–99.

Boulton, G. S. 1979. Processes of glacial erosion on different substrata. *Journal of Glaciology* **23**, 15–38.

Boulton, G. S. 1990. Sedimentary and sea level changes during glacial cycles and their control on glacimarine facies architecture. See Dowdeswell & Scourse (1990), 15–52.

Boulton, G. S. & R. C. A. Hindmarsh 1987. Sediment deformation beneath glaciers: rheology and geological consequences. *Journal of Geophysical Research* **92** (B9), 9059–82.

Boyd, R., D. B. Scott, M. Douna 1988. Glacial tunnel valleys and Quaternary history of the outer Scotian Shelf. *Nature* **333**, 61–4.

Bridle, I. M. & P. H. Robinson 1989. Diagenesis. See Barrett (1989b), 201–7.

Brodzikowski, K. & A. J. van Loon 1991. *Glacigenic sediments*. Amsterdam: Elsevier.

von Brunn, V. & T. Stratton 1981. Late Palaeozoic tillites of the Karoo Basin of South Africa. See Hambrey & Harland (1981), 71–9.

Bryant, A. D. 1983. The utilization of Arctic river analogue studies in the interpretation of periglacial river sediments in southern Britain. In *Background to palaeohydrology*, K. J. Gregory (ed.), 413–31. Chichester, England: John Wiley.

Budd, W. & B. J. McInnes 1978. Modelling surging glaciers and periodic surging of the Antarctic ice sheet. In *Climatic change and variability: a Southern Hemisphere perspective*, A. B. Pittock, L. A. Frakes, D. Jenssen, J. A. Peterson & J. W. Zillman (eds), 228–33. Cambridge: Cambridge University Press.

Budd, W. F., T. H. Jacka, V. I. Morgan 1980. Antarctic iceberg melt rates derived from size distributions and movement rates. *Annals of Glaciology* **1**, 103–12.

Budd, W., M. J. Corry, T. H. Jacka 1982. Results from the Amery Ice Shelf Project. *Annals of Glaciology* **3**, 36–41.

Cameron, D. 1965. Early discoveries XXII, Goethe – discoverer of the Ice Age. *Journal of Glaciology* **5**, 751–4.

Carey, S. W. & N. Ahmed 1961. Glacial marine sedimentation. In *Geology of the Arctic*, vol. 2, G. O. Raasch (ed.), 865–94. Toronto: University of Toronto Press.

Carlson, P. R., T. R. Bruns, M. A. Fisher 1990. Development of slope valleys in the glacimarine environ-ments of a complex subduction zone, Northern Gulf of Alaska. See Dowdeswell & Scourse (1990), 139–54.

Carol, H. 1947. The formation of roches moutonnées. *Journal of Glaciology* **1**, 57–9.

Chinn, T. J., M. J. McSaveney, E. R. McSaveney 1992. *The Mount Cook rock avalanche of 14 December 1991*. Information brochure. Wellington, New Zealand: Institute of Geological & Nuclear Sciences

Church, M. & R. Gilbert 1975. Proglacial fluvial and lacustrine environments. See Jopling & McDonald (1975), 155–76.

Clark, D. L. & A. Hanson 1983. Central Arctic Ocean sediment texture: a key to ice transport mechanisms. See Molnia (1983b), 301–30.

Clarke, G. K. C. 1987. A short history of scientific investigations on glaciers. In *Journal of Glaciology, Special Issue Commemorating Fiftieth Anniversary of the International Glaciological Society*, 4–24.

Clarke, G. K. C. 1991. Length, width and slope influences on glacier surging. *Journal of Glaciology* **37**, 236–46.

Clarke, G. K. C., S. G. Collins, D. E. Thompson 1984. Flow, thermal structure and subglacial conditions of a surge-type glacier. *Canadian Journal of Earth Sciences* **21**, 232–40.

Clarke, T. S. 1991. Glacier dynamics in the Susitna River basin, Alaska, USA. *Journal of Glaciology* **37**, 97–106.

Clayton, L., J. T. Teller, J. W. Attig 1985. Surging of the southwestern part of the Laurentide Ice Sheet. *Boreas* **14**, 235–41.

Clemmensen, L. B. & M. Houmark-Nielsen 1981. Sedimentary features of a Weichselian glaciolacustrine delta. *Boreas* **10**, 229–45.

Collins, D. N. 1977. Hydrology of an Alpine glacier as indicated by the chemical composition of meltwater. *Zeitschrift für Gletscherkunde und Glazialgeologie* **13**, 219–38.

Cooper, A. K., P. J. Barrett, K. Hinz, V. Traube, G. Leitchenkov, H. M. J. Stagg 1991a. Cenozoic prograding sequences of the Antarctic continental margin: a record of glacio-eustatic and tectonic events. *Marine Geology* **102**, 175–213.

Cooper, A. K., H. M. J. Stagg, E. Geist 1991b. Seismic stratigraphy and structure of Prydz Bay, Antarc-tica: implications from ODP Leg 119 drilling. See Barron et al. (1991), 5–25.

Cowan, E. A. & R. D. Powell 1990. Suspended sediment transport and deposition of cyclically interlaminated sediment in a temperate glacial fjord, Alaska, USA. See Dowdeswell & Scourse (1990), 75–90.

Crabtree, R. D. & C. S. M. Doake 1980. Flow lines on Antarctic ice shelves. *Polar Record* **20**, 31–7.

Croot, D. G. 1987. Glacio-tectonic structures: a mesoscale model of thin-skinned thrust sheets? *Journal of Structural Geology* **9**, 797–808.

Dahl, R. 1965. Plastically sculptured detail forms on rock surfaces in northern Nordland, Norway. *Geografiska Annaler* **47**, 83–140.

Deynoux, M. 1980. *Les formations glaciares du Précambrien terminal et de la fin de l'Ordovicien en Afrique de l'Ouest. Deux exemples de glaciation d'inlandsis sur une plate-forme stable*. Travaux des Laboratoires des Sciences de la Terre, St. Jerôme, Marseilles.

Deynoux, M. & R. Trompette 1981a. Late Ordovician tillites of the Taoudeni Basin, West Africa. See Hambrey & Harland (1981), 89–96.

Deynoux, M. & R. Trompette 1981b. Late Precambrian tillites of the Taoudeni Basin, West Africa. See Hambrey & Harland (1981), 123–31.

Doake, C. S. M. & D. G. Vaughan 1990. Rapid disintegration of the Wordie Ice Shelf in response to atmos-pheric warming. *Nature* **350**, 328–30.

Domack, E. W. 1982. Sedimentology of glacial and glacial marine deposits on the George V – Adélie conti-nental shelf, East Antarctica. *Boreas* **11**, 79–97.

Domack, E. W. 1988. Biogenic facies in the Antarctic glacimarine environment: basis for a polar glacimarine summary. *Palaeogeography, Palaeoclimatology, Palaeoecology* **63**, 357–72.

Domack, E. W. 1984. Rhythmically bedded glaciomarine sediments on Whidbey Island, northwestern Wash-ington State and southwestern British Columbia. *Journal of Sedimentary Petrology* **54**, 589–602.

Domack, E. W. 1990. Laminated terrigenous sediments from the Antarctic Peninsula: the role of subglacial and marine processes. See Dowdeswell & Scourse (1990), 91–104.

Domack, E. W. & D. E. Lawson 1985. Pebble fabric in an ice-rafted diamicton. *Journal of Geology* **93**,

577–92.

Domack, E. W., J. B. Anderson, D. D. Kurtz 1980. Clast shape as an indicator of transport and depositional mechanisms in glacial marine sediments: George V continental shelf, Antarctica. *Journal of Sedimentary Petrology* **50**, 813–20.

Dowdeswell, J. A. 1986. Distribution and character of sediments in a tidewater glacier, southern Baffin Island, N. W. T., Canada. *Arctic and Alpine Research* **18**, 45–56.

Dowdeswell, J. A. 1987. Processes of glacimarine sedimentation. *Progress in Physical Geography* **11**, 52–90.

Dowdeswell, J. A. & R. L. Collin 1990. Fast-flowing outlet glaciers on Svalbard ice caps. *Geology* **18**, 778–81.

Dowdeswell, J. A. & T. Murray 1990. Modelling rates of sedimentation from icebergs. See Dowdeswell & Scourse (1990), 121–38.

Dowdeswell, J. A. & J. C. Scourse (eds) 1990. *Glacimarine environments: processes and sediments*. London: Geological Society of London, Special Publication 53.

Dowdeswell, J. A. & M. J. Sharp 1986. The characterization of pebble fabrics in modern glacigenic sediments. *Sedimentology* **33**, 699–710.

Dowdeswell, J. A., M. J. Hambrey, R. T. Wu 1985. A comparison of clast fabric and shape in late Precambrian and modern glacigenic sediments. *Journal of Sedimentary Petrology* **55**, 691–704.

Dreimanis, A. 1976. Tills: their origin and properties. See Leggett (1976), 11–49.

Dreimanis, A. 1979. The problems of waterlain tills. In *Moraines and varves*, Ch. Schlüchter (ed.), 167–77. Rotterdam: Balkema.

Dreimanis, A. 1984. Lithofacies types and vertical profile models; an alternative approach to the description and environmental interpretation of glacial diamict and diamictite sequences. Discussion. *Sedimentology* **31**, 885–6.

Dreimanis, A. 1989. Tills: their genetic terminology and classification. See Goldthwait & Matsch (1989), 17–83.

Drewry, D. J. 1983. *Antarctica: Glaciological and geophysical folio*. Cambridge: Scott Polar Research Institute.

Drewry, D. J. 1986. *Glacial geologic processes*. London: Edward Arnold.

Drewry, D. J. 1991. The response of the Antarctic ice sheet to climatic change. In *Antarctica and global climate change*, C. Harris & B. Stonehouse (eds), 90–106. Cambridge: Scott Polar Research Institute & Belhaven Press.

Drewry, D. J. & A. P. R. Cooper 1981. Processes and models of Antarctic glaciomarine sedimentation. *Annals of Glaciology* **2**, 117–22.

Dunbar, R. B., J. B. Anderson, E. W. Domack 1985. Oceanographic influences on sedimentation along the Antarctic continental shelf. In *Oceanology of the Antarctic continental shelf*, 291–312. Antarctic Research Series 43.

Ehlers, J. (ed.) 1983. *Glacial deposits in North-West Europe*. Rotterdam: Balkema.

Ehlers, J., P. L. Gibbard, J. Rose (eds) 1991. *Glacial deposits in Great Britain and Ireland*. Rotterdam: Balkema.

Ehrmann, W. U. 1991. Implications of sediment composition in the southern Kerguelen Plateau for paleoclimate and depositional environment. See Barron et al. (1991), 185–210.

Ehrmann, W. U. & A. Mackensen 1992. Sedimentological evidence for the formation of an East Antarctic ice sheet in Eocene/Oligocene time. *Palaeogeography, Palaeoclimatology, Palaeoecology* **93**, 85–112.

Elson, J. A. 1989. Comment on glacitectonite, deformation till and comminution till. See Goldthwait & Matsch (1989), 85–8.

Elverhøi, A. 1984. Glacigenic and associated marine sediments in the Weddell Sea, fjords of Spitsbergen and the Barents Sea: a review. *Marine Geology* **57**, 53–88.

Elverhøi, A. & E. Roaldset 1983. Glaciomarine sediments and suspended particulate matter, Weddell Sea shelf, Antarctica. *Polar Research* **1**, 1–21.

Elverhøi, A., O. Lonne, R. Seland 1983. Glaciomarine sedimentation in a modern fjord environment, Spitsbergen. *Polar Research* **1**, 127–49.

Embleton, C. & C. A. M. King 1968. *Glacial and periglacial geomorphology*. London: Edward Arnold.

Embleton, C. & C. A. M. King 1975. *Glacial geomorphology*. London: Edward Arnold.

Eyles, C. H. 1988a. Glacially and tidally-influenced shallow marine sedimentation of the late Precambrian

Port Askaig Formation. *Palaeogeography, Palaeoclimatology, Palaeoecology* **68**, 1–25.

Eyles, C. H. 1988b. A model for striated boulder pavement formation on glaciated shallow marine shelves, an example from the Yakataga Formation, Alaska. *Journal of Sedimentary Petrology* **58**, 62–71.

Eyles, C. H. & N. Eyles 1983. Sedimentation in a large lake: a reinterpretation of the late Pleistocene stratigraphy at Scarborough Bluffs, Ontario, Canada. *Geology* **11**, 146–52. .

Eyles, C. H. & M. B. Lagoe 1990. Sedimentation patterns and facies geometries on a temperate glacially-influenced continental shelf: the Yakataga Formation, Middleton Island, Alaska. See Dowdeswell & Scourse (1990), 363–86.

Eyles, N. (ed.) 1983. *Glacial geology*. Oxford: Pergamon.

Eyles, N. & M. B. Lagoe 1989. Sedimentology of shell-rich deposits (coquinas) in the glaciomarine upper Cenozoic Yakataga Formation, Middleton Island, Alaska. *Geological Society of America, Bulletin* **101**, 129–42.

Eyles, N. & A. M. McCabe 1989. Glaciomarine facies within subglacial tunnel valleys: the sedimentary record of glacio-isostatic downwarping in the Irish Sea Basin. *Sedimentology* **36**, 431–48.

Eyles, N. & J. Menzies 1983. The subglacial landsystem. See Eyles (1983), 19–70.

Eyles, N. & A. D. Miall 1984. Glacial facies models. In *Facies models*, R. G. Walker (ed.), 15–38. Toronto: Geological Association of Canada.

Eyles, N., C. H. Eyles, A. D. Miall 1983. Lithofacies types and vertical profile models; an alternative approach to the description and environmental interpretation of glacial diamict and diamictite sequences. *Sedimentology* **30**, 393–410.

Fairchild, I. J. 1983. Effects of glacial transport and neomorphism on Precambrian dolomite crystal sizes. *Nature* **304**, 714–16.

Fairchild, I. J. 1993. Balmy shores and icy wastes: the paradox of carbonates associated with glacial deposits in Neoproterozoic times. *Sedimentology Review* **1**, 1–16.

Fairchild, I. J. & M. J. Hambrey 1984. The Vendian succession of north-eastern Spitsbergen: petrogenesis of a dolomite–tillite association. *Precambrian Research* **26**, 111–67.

Fairchild, I. J. & M. J. Hambrey 1994. Vendian basin evolution in East Greenland and NE Svalbard. *Precambrian Research*, in press.

Fairchild, I. J. & B. Spiro 1990. Carbonate minerals in glacial sediments: geochemical clues to palaeoenvironment. See Dowdeswell & Scourse (1990), 241–56.

Fairchild, I. J., G. L. Hendry, M. Quest, M. E. Tucker 1988. Chemical analysis of sedimentary rocks. In *Techniques in sedimentology*, M. E. Tucker (ed.), 274–354. Oxford: Blackwell Scientific.

Fairchild, I. J., M. J. Hambrey, B. Spiro, T. H. Jefferson 1989. Late Proterozoic glacial carbonates in NE Spitsbergen: new insights into the carbonate–tillite association. *Geological Magazine* **126**, 469–90.

Fairchild, I. J., L. Bradby, B. Spiro 1993. Reactive carbonate in glacial sediments: a preliminary synthesis of its creation, dissolution and reincarnation. In *Earth's glacial record*, M. Deynoux et al. (eds). Cambridge: Cambridge University Press.

Flint, R. F. 1971. *Glacial and Quaternary geology*. New York: John Wiley.

Flint, R. F., J. E. Sanders, J. Rodgers 1960. Diamictite: a substitute term for symmictite. *Geological Society of America, Bulletin* **71**, 1809–10.

Folk, R. L. 1975. Glacial deposits identified by chattermark trails in detrital garnets. *Geology* **3**, 473–5.

Ford, D. C., P. G. Fuller, J. J. Drake 1970. Calcite precipitates at the soles of temperate glaciers. *Nature* **226**, 441–2.

Frakes, L. A. 1975. Geochemistry of Ross Sea diamicts. In *Initial Reports Deep Sea Drilling Project, Leg 28*, D. E. Hayes, L. A. Frakes & Shipboard Scientific Party (eds), 789–94. Washington, DC: US Government Printing Office.

Frakes, L. A. 1979. *Climates throughout geological time*. Amsterdam: Elsevier.

Frakes, L. A. & J. E. Francis 1988. A guide to Phanerozoic cold polar climates from high-latitude ice-rafting in the Cretaceous. *Nature* **333**, 547–9.

Frakes, L. A., J. E. Francis, J. I. Sykes 1992. *Climate modes of the Phanerozoic*. Cambridge: Cambridge University Press.

Francis, E. 1975. Glacial sediments: a selective review. In *Ice ages: ancient and modern*, A. E. Wright & F. Moseley (eds), 43–68. Liverpool: Seel House Press.

285

Fütterer, D. K. & M. Melles 1990. Sediment patterns in the southern Weddell Sea: Filchner Shelf and Filchner Depression. In *Geological history of the polar oceans: Arctic versus Antarctic*, U. Bleil & J. Theide (eds), 381–401. Dordrecht: Kluwer Academic Publishers.

Fyfe, G. J. 1990. The effect of water depth on ice-proximal glaciolacustrine sedimentation: Salpausselkè I, southern Finland. *Boreas* **19**, 147–64.

Garwood, E. J. 1932. Speculation and research in Alpine glaciology: an historical review. *Quarterly Journal of the Geological Society of London* **88**, xciii–cxviii.

Glen, J. W. 1952. Experiments on the deformation of ice. *Journal of Glaciology* **2**, 111–14.

Goldthwait, R. P. 1951. Development of end moraines in east-central Baffin Island. *Journal of Geology* **59**, 567–77.

Goldthwait, R. P. 1971. Introduction to till, today. In *Till: a symposium*, R. P. Goldthwait (ed.), 3–26. Columbus: Ohio State University Press.

Goldthwait, R. P. 1979. Giant grooves made by concentrated basal ice streams. *Journal of Glaciology* **23**, 297–307.

Goldthwait, R. P. 1989. Classification of glacial morphologic features. See Goldthwait & Matsch (1989), 267–77.

Goldthwait, R. P. & C. L. Matsch (eds) 1989. *Genetic classification of glacigenic deposits*. Rotterdam: Balkema.

Goodwin, R. G. 1984. *Neoglacial lacustrine sedimentation and ice advance, Glacier Bay, Alaska*. Ohio State University, Institute of Polar Studies, Report 79.

Gravenor, C. P. 1979. The nature of the Late Palaeozoic glaciation in Gondwana as determined from an analysis of garnets and other heavy minerals. *Canadian Journal of Earth Sciences* **16**, 1137–53.

Gravenor, C. P. 1980. Heavy minerals and sedimentological studies on the glaciogenic Late Precambrian Gaskiers Formation of Newfoundland. *Canadian Journal of Earth Sciences* **17**, 1131–41.

Gravenor, C. P. 1981. Chattermark trails on garnets as an indicator of glaciation in ancient deposits. See Hambrey & Harland (1981), 17–18.

Griffith, T. W. & J. B. Anderson 1989. Climatic control of sedimentation in bays and fjords of the northern Antarctic Peninsula. *Marine Geology* **85**, 181–204.

Gripp, K. 1929. Glaciologische unde geologische Ergebnisse der Hamburgischen Spitzbergen-Expedition 1927. *Abhandlungen aus dem Gebeite der Naturwissenschaften, Herausgegeben vom Naturwissenschaftlichen Verien in Hamburg.* **22** (2–4), 146–249.

Grove, J. M. 1988. *The Little Ice Age*. London: Methuen.

Gustavson, T. C., G. M. Ashley, J. C. Boothroyd 1975. Depositional sequences in glaciolacustrine deltas. See Jopling & McDonald (1975), 264–80.

Haase, G. H. 1986. Glaciomarine sediments along the Filchner/Ronne Ice Shelf, southern Weddell Sea – first results of the 1983/84 Antarktis-II/4 Expedition. *Marine Geology* **72**, 241–58.

Hall, K. J. 1989. Clast shape. See Barrett (1989b), 63–6.

Hallet, B. 1976. The effect of subglacial chemical processes on glacier sliding. *Journal of Glaciology* **17**, 209–21.

Hambrey, M. J. 1975. The origin of foliation in glaciers: evidence from some Norwegian examples. *Journal of Glaciology* **14**, 181–5.

Hambrey, M. J. 1977. Foliation, minor folds and strain in glacier ice. *Tectonophysics* **39**, 397–416.

Hambrey, M. J. 1982. Late Proterozoic diamictites of northeastern Svalbard. *Geological Magazine* **119**, 527–51.

Hambrey, M. J. 1989. Grain fabric studies on the CIROS-1 core. See Barrett (1989b), 59–62.

Hambrey, M. J. 1991. Structure and dynamics of the Lambert Glacier–Amery Ice Shelf system: implications for the origin of Prydz Bay sediments. See Barron et al. (1991), 61–76.

Hambrey, M. J. 1992. Secrets of a tropical ice age. *New Scientist* (1 February), 42–9.

Hambrey, M. J. & J. C. Alean 1992. *Glaciers*. Cambridge: Cambridge University Press.

Hambrey, M. J. & J. A. Dowdeswell 1994. Flow regime of the Lambert Glacier–Amery Ice Shelf system, Antarctica: structural evidence from satellite imagery. *Annals of Glaciology*, in press.

Hambrey, M. J. & W. B. Harland (collators and eds) 1981. *Earth's pre-Pleistocene glacial record*. Cambridge: Cambridge University Press.

Hambrey, M. J. & A. G. Milnes 1975. Boudinage in glacier ice – some examples. *Journal of Glaciology* **14**,

383–93.

Hambrey, M. J. & Milnes, A. G. 1977. Structural geology of an Alpine glacier (Griesgletscher, Valais, Switzerland). *Eclogae Geologicae Helvetiae* **70**, 667–84.

Hambrey, M. J. & Müller, F. 1978. Ice deformation and structures in the White Glacier, Axel Heiberg Island, Northwest Territories, Canada. *Journal of Glaciology* **20**, 41–66.

Hambrey, M. J., A. G. Milnes, H. Siegenthaler 1980. Dynamics and structure of Griesgletscher, Switzerland. *Journal of Glaciology* **25**, 215–28.

Hambrey, M. J., P. J. Barrett, K. J. Hall, P. H. Robinson 1989. Stratigraphy. See Barrett (1989b), 23–48.

Hambrey, M. J., W. U. Ehrmann, B. Larsen 1991. The Cenozoic glacial record of the Prydz Bay continental shelf, East Antarctica. See Barron et al. (1991), 77–132.

Hambrey, M. J., P. J. Barrett, W. U. Ehrmann, B. Larsen 1992. Cenozoic sedimentary processes on the Antarctic continental shelf: the record from deep drilling. *Zeitschrift fur Geomorphologie* (Suppl. -Bd) **86**, 77–103.

Haq, B. U., J. Hardenbol, P. R. Vail 1987. Chronology of fluctuating sea levels since the Triassic. *Science* **235**, 1156–67.

Harland, W. B. 1957. Exfoliation joints and ice action. *Journal of Glaciology* **3**, 8–10.

Harland, W. B. 1964. Critical evidence for a great Infra-Cambrian glaciation. *Geologische Rundschau* **54**, 45–61.

Harland, W. B. & K. N. Herod 1975. Glaciations through time. In *Ice ages: ancient and modern*, A. E. Wright & F. Moseley (eds), 189–216. Liverpool: Seel House Press.

Harland, W. B., K. N. Herod, D. H. Krinsley 1966. The definition and identification of tills and tillites. *Earth Science Reviews* **2**, 225–56.

Harwood, D. M., P. J. Barrett, A. R. Edwards, H. J. Rieck, P-N. Webb 1989. Biostratigraphy and chronology. See Barrett (1989b), 231–9.

Hayes, D. E., L. A. Frakes, Shipboard Scientific Party 1975. *Initial Reports of the Deep Sea Drilling Project, Leg 28*. Washington, DC: US Government Printing Office.

Haynes, V. M. 1968. The influence of glacial erosion and rock structure on corries in Scotland. *Geografiska Annaler* **50A**, 221–34.

Hinz, K. & M. Block 1983. Results of geophysical investigations in the Weddell Sea and in the Ross Sea, Antarctica. In *Proceedings of the 11th World Petroleum Congress*, 75–91. London & New York: John Wiley.

Hinz, K. & Y. Kristoffersen 1987. Antarctica: recent advance in the understanding of the continental shelf. *Geologische Jarbuch* **E37**, 3–54.

Hooke, R. LeB. 1968. Comments on the formation of shear moraines: an example from south Victoria Land, Antarctica. *Journal of Glaciology* **7**, 351–2.

Hooke, R. LeB. & P. J. Hudleston 1978. Origin of foliation in glaciers. *Journal of Glaciology* **20**, 285–99.

Hooke, R. LeB., B. B. Dahlin, M. T. Kauper 1972. Creep of ice containing dispersed fine sand. *Journal of Glaciology* **11**, 327–36.

Hoppe, G. 1957. Problems of glacial geomorphology and the ice age. *Geografiska Annaler* **39**, 1–17.

Hudleston, P. J. 1976. Recumbent folding in the base of the Barnes Ice Cap, Baffin Island, Northwest Territories, Canada. *Geological Society of America, Bulletin* **87**, 1684–92.

Hudleston, P. J. & R. LeB. Hooke 1980. Cumulative deformation in the Barnes Ice Cap, and implications for the development of foliation. *Tectonophysics* **66**, 127–46.

Hughes, T. J. 1975. The West Antarctic Ice Sheet: instability, disintegration and the initiation of ice ages. *Review of Geophysics and Space Physics* **15**, 1–46.

Imbrie, J. & K. P. Imbrie 1979. *Ice ages: solving the mystery*. London: Macmillan.

IUGS 1989. Global stratigraphic chart. *Episodes* **12** (2), insert.

Jopling, A. V. & B. C. McDonald (eds) 1975. *Glaciofluvial and glaciolacustrine sedimentation*. Tulsa: Society of Economic Paleontologists and Mineralogists, Special Publication 23.

Kälin, M. 1971. The active push moraine of the Thomson Glacier. In *Axel Heiberg Island Research Report, Glaciology* 4. McGill University, Montreal.

Kamb, W. B. 1987. Glacier surge mechanism based on linked cavity configuration of the basal water conduit system. *Journal of Geophysical Research* **92** (B9), 9083–100.

Kamb, W. B., C. F. Raymond, W. D. Harrison, H. F. Engelhardt, K. Echelmeyer, N. Humphry, M. Brugman, T. Pfeffer 1985. Glacier surge mechanism: the 1982–1983 surge of Variegated Glacier, Alaska,

Science **227**, 469–79.

Karrow, P. F. 1984. Lithofacies types and vertical profile models; an alternative approach to the description and environmental interpretation of glacial diamict and diamictite sequences. Discussion. *Sedimentology* **31**, 883–4.

Kemmis, T. J. & T. J. Hallberg 1984. Lithofacies types and vertical profile models; an alternative approach to the description and environmental interpretation of glacial diamict and diamictite sequences. Discussion. *Sedimentology* **31**, 886–90.

King, L. H. 1994. Till in the marine environment. *Journal of Quaternary Science* **8**, 347–58.

King, L. H. & G. Fader 1987. Wisconsinan glaciation on the continental shelf – southeast Atlantic Canada. *Geological Survey of Canada, Bulletin* **363**, 1–72.

King, L. H., K. Rokoengen, T. Gunleiksrud 1987. *Quaternary seismostratigraphy of the Mid Norwegian shelf, 65°–67°30'N – a till tongue stratigraphy*. IKU, Publication 114.

King, L. H., K. Kokoengen, G. B. J. Fader, T. Gunleiksrud 1991. Till-tongue stratigraphy. *Geological Society of America, Bulletin* **103**, 637–59.

Krinsley, D. H. & J. C. Doornkamp 1973. *Atlas of quartz and sand surface textures*. Cambridge: Cambridge University Press.

Krinsley, D. H. & B. Funnell 1965. Environmental history of sand grains from the Lower and Middle Pleistocene of Norfolk, England. *Quarterly Journal of the Geological Society of London* **121**, 435–56.

Krumbein, W. C. 1941. Measurement and geological significance of shape and roundness of sedimentary particles. *Journal of Sedimentary Petrology* **11**, 64–72.

Kuhn, G., M. Melles, W. U. Ehrmann, M. J. Hambrey, G. Schmiedl 1993. Character of clasts in glaciomarine sediments as an indicator of transport and depositional processes, Weddell and Lazarev Seas, Antarctica. *Journal of Sedimentary Petrology* **63**, 477–87.

Kujansuu, R. 1976. Glaciogeological surveys for ore-prospecting purposes in northern Finland. See Leggett (1976), 225–39.

Lange, M. A. & D. R. MacAyeal 1988. Numerical models of steady-state thickness and basal ice configurations of the central Ronne Ice Shelf, Antarctica. *Annals of Glaciology* **11**, 64–70.

Larter, R. D. & P. F. Barker 1989. Seismic stratigraphy of the Antarctic Peninsula Pacific margin: a record of Pliocene–Pleistocene ice volume and paleoclimate. *Geology* **17**, 731–4.

Lawson, D. E. 1979. A comparison of the pebble orientations in ice and deposits of the Matanuska Glacier, Alaska. *Journal f Geology* **87**, 629–45.

Lawson, D. E. 1981. *Sedimentological characteristics and classification of depositional processes and deposits in the glacial environment*. US Army Cold Regions Research and Engineering Laboratory, Report 81–27.

Lawson, D. E. 1982. Mobilization, movement and deposition of active subaerial sediment flows, Matanuska Glacier, Alaska. *Journal of Geology* **90**, 279–300.

Leggett, R. F. (ed.) 1976. *Glacial till*. Royal Society of Canada, Special Publication 12.

Lewis, W. V. 1954. Pressure release and glacial erosion. *Journal of Glaciology* **2**, 417–22.

Lewis, W. V. 1960. *Norwegian cirque glaciers*. Royal Geographical Society, Research Series 4.

Lien, R., A. Solheim, A. Elverhøi, K. Rokøngin 1989. Iceberg scouring and sea bed morphology on the eastern Weddell Sea Shelf. *Polar Research* **4**, 43–57.

Lundqvist, J. 1989. Rogen (ribbed) moraine – identification and possible origin. *Sedimentary Geology* **62**, 281–92.

Mackiewicz, N. E., R. D. Powell, P. R. Carlson, B. F. Molnia 1984. Interlaminated ice-proximal glacimarine sediments in Muir Inlet, Alaska. *Marine Geology* **57**, 113–47.

Manley, G. 1959. The late-glacial climate of North-west England. *Liverpool and Manchester Geological Journal* **2**, 188–215.

May, R. W. & A. Dreimanis 1976. Compositional variability in tills. See Leggett (1976), 99–120.

McCabe, A. M. & N. Eyles 1988. Sedimentology of an ice-contact glaciomarine delta, Carey Valley, Northern Ireland. *Sedimentary Geology* **59**, 1–14.

McDonald, B. C. & W. W. Shilts 1975. Interpretation of faults in glaciofluvial sediments. See Jopling & McDonald (1975), 123–31.

McIntyre, N. F. 1985. A re-assessment of the mass balance of the Lambert Glacier drainage basin, Antarctica. *Journal of Glaciology* **31**, 34–8.

Meier, M. F. 1960. *Mode of flow of Saskatchewan Glacier, Alberta, Canada*. US Geological Survey, Professional Paper 351.

Menzies, J. 1989. Subglacial hydraulic conditions and their possible impact upon subglacial bed formation. *Sedimentary Geology* **62**, 125–50.

Menzies, J. & J. Rose (eds) 1987. *Drumlin Symposium*. Rotterdam: Balkema.

Menzies, J. & J. Rose 1989. Subglacial bedforms – drumlins, Rogen moraine and associated subglacial bedforms. *Sedimentary Geology, Special Issue* **62** (2–4).

Miall, A. D. 1977. A review of the braided-river depositional environment. *Earth Science Reviews* **13**, 1–62.

Miall, A. D. 1978. Lithofacies types and vertical profile models in braided river deposits: a summary. In *Fluvial sedimentology*, A. D. Miall (ed.), 597–604. Canadian Society of Petroleum Geology, Memoir 5.

Miall, A. D. 1983a. Glaciomarine sedimentation in the Gowganda Formation (Huronian), Northern Ontario. *Journal of Sedimentary Petrology* **53**, 477–91.

Miall, A. D. 1983b. Glaciofluvial transport and deposition. See Eyles (1983), 168–83.

Mickelsen, D. M. 1971. *Glacial geology of the Burroughs Glacier area, southeast Alaska*. Columbus: Ohio State University, Institute of Polar Studies, Report 40.

Miller, K. G., R. G. Fairbanks, G. S. Mountain 1987. Tertiary oxygen isotope synthesis, sea level history, and continental margin erosion. *Paleooceanography* **2**, 1–19.

Mills, H. H. 1977. Basal till fabrics of modern alpine glacier. *Geological Society of America, Bulletin* **88**, 824–8.

Mills, W. 1983. Darwin and the iceberg theory. *Notes and Records of the Royal Society of London* **38**, 109–27.

Molnia, B. F. 1983a. Subarctic glacial-marine sedimentation: a model. See Molnia (1983b), 95–144.

Molnia, B. F. (ed.) 1983b. *Glacial-marine sedimentation*. New York & London: Plenum Press.

Moncrieff, A. C. M. 1989. Classification of poorly sorted sedimentary rocks, *Sedimentary Geology* **65**, 191–4.

Moncrieff, A. C. M. & M. J. Hambrey 1988. Late Precambrian glacially-related grooved and striated surfaces in the Tillite Group of central East Greenland. *Palaeogeography, Palaeoclimatology & Palaeoecology* **65**, 183–200.

Morgan, V. I. 1972. Oxygen isotope evidence for bottom freezing on the Amery Ice Shelf. *Nature* **238**, 393–4.

Mustard, P. S. & J. A. Donaldson 1987. Early Proterozoic ice-proximal glaciomarine deposition: The lower Gowganda Formation at Cobalt, Ontario, Canada. *Geological Society of America, Bulletin* **98**, 373–87.

Nesbitt, H. W. & G. M. Young 1982. Early Proterozoic climates and plate motions inferred from major element chemistry of lutites. *Nature* **299**, 715–17.

Nye, J. F. 1952 The mechanics of glacier flow. *Journal of Glaciology* **2**, 82–93.

Nye, J. F. 1957. The distribution of stress and velocity in glaciers and ice sheets. *Royal Society of London, Proceedings* **239A**, 113–33.

Nye, J. F. 1958. *A theory of wave formation on glaciers*, 139–54. International Association of Hydrological Sciences, Publication 47.

Nye, J. F. 1973. *Water at the bed of a glacier*, 189–94. International Association of Hydrological Sciences, Publication 95.

Nye, J. F. & F. C. Frank 1973. *Hydrology of the intergranular veins in a temperate glacier*, 157–61. International Association of Hydrological Sciences, Publication 95.

Orheim, O. & A. Elverhøi 1981. Model for submarine glacial deposition. *Annals of Glaciology* **2**, 123–9.

Østrem, G. 1975. Sediment transport in glacial streams. See Jopling & McDonald (1975), 101–22.

Oswald, G. K. A. & G. de Q. Robin 1973. Lakes beneath the Antarctic ice sheet. *Nature* **245**, 251–4.

Parker, B. C., G. M. Simmons, F. G. Love, R. A. Wharton, K. G. Seaburg 1981. Modern stromatolites in Antarctic Dry Valley lakes. *Bioscience* **31**, 656–61.   Paterson, W. S. B. 1981. *The physics of glaciers*. Oxford: Pergamon.

Paul, M. A. 1983. The supraglacial landsystem. See Eyles (1983), 71–90.

Peacock, J. D. & Cornish, R. 1989. *Glen Roy area*. Cambridge: Quaternary Research Association Field Guide.

Pettijohn, F. J. 1975. *Sedimentary rocks*. New York: Harper & Row.

Phillips, F. C. 1971. *The use of stereographic projection in structural geology*. London: Edward Arnold.

Post, A. & E. R. La Chapelle 1971. *Glacier ice*. Seattle: University of Washington Press and the Mountaineers.

Powell, R. D. 1981. A model for sedimentation by tidewater glaciers. *Annals of Glaciology* **2**, 129–34.

Powell, R. D. 1983. Glacial-marine sedimentation processes and lithofacies of temperate tidewater glaciers,

Glacier Bay, Alaska. See Molnia (1983b), 185–232.

Powell, R. D. 1984. Glacimarine processes and inductive lithofacies modelling of ice shelf and tidewater glacial sediments based on Quaternary examples. *Marine Geology* **57**, 1–52.

Powell, R. D. 1990. Glacimarine processes at grounding-line fans and their growth to ice-contact deltas. See Dowdeswell & Scourse (1990), 53–74.

Powell, R. D. & B. F. Molnia 1989. Glacimarine sedimentation processes, facies and morphology of the south-southeast Alaska shelf and fjords. *Marine Geology* **85**, 359–90.

Prest, V. K. 1983. *Canada's heritage of glacial features*. Geological Survey of Canada, Miscellaneous Report 28.

Price, R. J. 1973. *Glacial and fluvioglacial landforms*. Edinburgh: Oliver & Boyd.

Ramsay, J. G. 1967. *Folding and fracturing of rocks*. New York: McGraw-Hill.

Raymond, C. F. 1987. How do glaciers surge? A review. *Journal of Geophysical Research* **92** (B9), 9121–34.

Reading, H. G. 1978. Facies. In *Sedimentary environments and facies*, H. G. Reading (ed.), 4–14. Oxford: Blackwell Scientific.

Reynolds, J. M. & M. J. Hambrey 1988. The structural glaciology of the George VI Ice Shelf, Antarctica. *British Antarctic Survey Bulletin* **79**, 79–95.

Robin, G. de Q. 1974. Depth of water-filled crevasses that are closely spaced (letter). *Journal of Glaciology* **13**, 543.

Robin, G. de Q. 1975. Ice shelves and ice flow. *Nature* **253**, 168–72.

Robin, G. de Q. 1979. Formation, flow and disintegration of ice shelves. *Journal of Glaciology* **24**, 259–71.

Robin, G. de Q. 1983. Coastal sites, Antarctica. In *Climatic record in polar ice sheets*, G. de Q. Robin (ed.), 118–22. Cambridge: Cambridge University Press.

Rose, J. 1987. Drumlins as part of a glacier bedform continuum. In *Drumlin Symposium*, J. Menzies & J. Rose (eds), 103–16. Rotterdam: Balkema.

Rose, J. 1989. Glacier stress patterns and sediment transfer associated with the formation of superimposed flutes. *Sedimentary Geology* **62**, 151–76.

Röthlisberger, H. 1972. Water pressure in intra- and subglacial channels. *Journal of Glaciology* **11**, 177–203.

Rust, B. R. 1975. Late Quaternary subaqueous outwash deposits near Ottawa, Canada. See Jopling & McDonald (1975), 177–92.

Sanderson, H. C. 1975. Sedimentology of the Brampton esker and its associated deposits: an empirical test of theory. See Jopling & McDonald (1975), 155–76.

Schlüchter, Ch. (ed.) 1979. *Moraine and varves*. Rotterdam: Balkema.

Schermerhorn, L. F. G. 1974. Late Precambrian mixtites: glacial and/or nonglacial? *American Journal of Science* **274**, 673–824.

Sexton, D. J., J. A. Dowdeswell, A. Solheim, A. Elverhøi 1992. Seismic architecture and sedimentation in northwest Spitsbergen fjords. *Marine Geology* **103**, 53–68.

Shabtaie, S. & C. R. Bentley 1987. West Antarctic ice streams draining into the Ross Ice Shelf: configuration and mass balance. *Journal of Geophysical Research* **92** (B9), 8865–84.

Shackleton, N. J. & J. K. Kennett 1975. Paleotemperature history of the Cenozoic and the initiation of Antarctic glaciation: oxygen and carbon isotope analyses in DSDP Sites 277, 279 and 281. In *Initial Reports of the Deep Sea Drilling Project, Leg 29*, J. P. Kennett, R. E. Houtz & Shipboard Scientific Party (eds), 743–55. Washington, DC: US Government Printing Office.

Sharp, M. 1982. Modification of clasts in lodgement tills by glacial erosion. *Journal of Glaciology* **28**, 475–81.

Sharp, M. 1985a. Crevasse-fill ridges: a landform type characteristic of surging glaciers? *Geografiska Annaler* **67A**, 213–20.

Sharp, M. 1985b. Sedimentation and stratigraphy of Eyabakkajükull – an Icelandic surging glacier. *Quaternary Research* **24**, 268–84.

Sharp, M. 1988a. Surging glaciers: behaviour and mechanisms. *Progress in Physical Geography* **12**, 349–70.

Sharp, M. 1988b. Surging glaciers: geomorphic effects. *Progress in Physical Geography* **12**, 533–59.

Sharp, M., J. C. Gemmell, J-L. Tison 1989. Structure and stability of the former subglacial drainage system of the Glacier de Tsanfleuron, Switzerland. *Earth Surface Processes and Landforms* **14**, 119–34.

Sharp, R. P. 1988. *Living ice: understanding glaciers and glaciation*. Cambridge: Cambridge University Press.

Shaw, J. 1975. Sedimentary successions in Pleistocene ice-marginal lakes. See Jopling & McDonald (1975), 281–303.

Shaw, J. 1977. Till deposited in arid polar environments. *Canadian Journal of Earth Sciences* **14**, 1239–45.

Shaw, J. 1989. Sublimation till. See Goldthwait & Matsch (1989), 141–2.

Shaw, J., D. Kvill, B. Rains 1989. Drumlins and catastrophic subglacial floods. *Sedimentary Geology* **62**, 177–202.

Shilts, W. W. 1976. Glacial till and mineral exploration. See Leggett (1976), 205–24.

Sladen, J. A. & W. Wrigley 1983. Geotechnical properties of lodgement till – a review. See Eyles (1983), 184–212.

Sissons, J. B. 1979. Catastrophic lake drainage in Glen Spean and the Great Glen, Scotland. *Geological Society of London, Journal* **136**, 215–24.

Solheim, A. 1991. *The depositional environment of surging sub-polar tidewater glaciers*. Skrifter Norsk Polarinstitutt 194.

Solheim, A. & Y. Kristofferson 1984. The physical environment, western Barents Sea, 1:1,500,000, Sheet B; Sediments above the upper regional unconformity; thickness, seismic structure and outline of the glacial history. *Norsk Polarinstitutt Skrifter* **179B**.

Solheim, A., L. Russwurm, A. Elverhøi, M. Nyland Berg 1990. Glacial geomorphic features in the northern Barents Sea: direct evidence for grounded ice and implications for the pattern of deglaciation and late glacial sedimentation. See Dowdeswell & Scourse (1990), 253–68.

Souchez, R. A. & M. Lemmens 1985. Subglacial carbonate deposition: an isotopic study of a present-day case. *Palaeogeography, Palaeoclimatology, Palaeoecology* **51**, 357–64.

Souchez, R. A. & R. D. Lorrain 1991. *Ice composition and glacier dynamics*. Berlin: Springer.

Spencer, A. M. 1971. *Late Precambrian glaciation in Scotland*. Geological Society of London, Memoir 6.

Spencer, A. M. 1985. Mechanisms and environments of deposition of the late Precambrian geosynclinal tillites: Scotland and East Greenland. *Palaeogeography, Palaeoclimatology, Palaeoecology* **51**, 143–57.

Srivastava, S. P., M. Arthur, Shipboard Scientific Party 1987. *Proceedings of the Ocean Drilling Program, Initial Reports 105*. College Station, Tex: Ocean Drilling Program.

Stagg, H. M. J. 1985. The structure and origin of Prydz Bay and MacRobertson Shelf, East Antarctica. *Tectonophysics* **114**, 315–40.

Stoker, M. S. 1990. Glacially-influenced sedimentation on the Hebridean slope, northwestern United Kingdom continental margin. See Dowdeswell & Scourse (1990), 349–62.

Stoker, M. S. & R. Holmes 1991. Submarine end-moraines as indicators of Pleistocene ice-limits off northwest Britain. *Journal of the Geological Society* **148**, 431–4.

Sturm, M. 1979. Origin and composition of varves. In *Moraines and varves*, Ch. Schlüchter (ed.), 281–5. Rotterdam: Balkema.

Sugden, D. & B. S. John 1976. *Glaciers and landscape*. London: Edward Arnold. Swithinbank, C. W. M. 1988. Antarctica. In *Satellite image atlas of the glaciers of the world*, R. S. Williams & J. G. Ferrigno (eds). US Geological Survey, Professional Paper 1386-B.

Syvitski, J. P. M. 1989. On the deposition of sediment within glacier-influenced fjords. *Marine Geology* **85**, 301–29.

Teller, J. T. 1985. Glacial Lake Agassiz and its influence on the Great Lakes. In *Quaternary evolution of the Great Lakes*, P. F. Karrow & P. E. Calkin (eds), 1–16. Geological Association of Canada, Special Paper 30.

Theakstone, W. H. 1967. Basal sliding and movement near the margins of the glacier Østerdalsisen, Norway. *Journal of Glaciology* **6**, 805–16.

Theakstone, W. H. 1976. Glacial lake sedimentation, Austerdalsisen, Norway. *Sedimentology* **23**, 671–88.

Theakstone, W. H. 1978. The 1977 drainage of Austre Okstindbreen ice-dammed lake, its causes and consequences. *Norsk Geografisk Tidsskrift* **32**, 159–71.

Thyssen, F. 1988. Special aspects of the central part of the Filchner–Ronne Ice Shelf, Antarctica. *Annals of Glaciology* **11**, 173–9.

Turner, B. R. 1991. Depositional environment and petrography of preglacial continental sediments from Hole 740A, Prydz Bay, Antarctica. See Barron et al. (1991), 45–56.

US National Park Service 1983. *Glacier Bay: a guide to Glacier Bay National Park and Preserve, Alaska*. Washington, DC: US National Park Service, Division of Publications.

Van der Meer, J. J. M. 1987. *Tills and glaciotectonics*. Rotterdam: Balkema.

291

Visser, J. N. J. 1983a. Glacial-marine sedimentation in the Late Paleozoic Karoo Basin, Southern Africa. See Molnia (1983b), 667–701.

Visser, J. N. J. 1983b. Submarine debris flow deposits from the Upper Carboniferous Dwyka Tille Formation in the Kalahari Basin, South Africa. *Sedimentology* **30**, 511–24.

Visser, J. N. J. & K. J. Hall 1985. Boulder beds in the glaciogenic Permo-Carboniferous Dwyka Formation in South Africa. *Sedimentology* **32**, 281–94.

Visser, J. N. J., J. C. Loock, W. P. Colliston 1987. Subaqueous outwash fan and esker sandstones in the Permo-Carboniferous Dwyka Formation of South Africa. *Journal of Sedimentary Petrology* **57**, 467–78.

Vivian, R. & G. Bocquet 1973. Subglacial cavitation phenomena under the Glacier d'Argentière, Mont Blanc, France. *Journal of Glaciology* **12**, 439–51.

Vorren, T. O., E. Lebesbye, K. Andreassen, K-B. Larsen 1989. Glacigenic sediments on a passive continental margin as exemplified by the Barents Sea. *Marine Geology* **85**, 251–72.

Vorren, T. O., E. Lebesbye, K. B. Larsen 1990. Geometry and genesis of the glacigenic sediments in the southern Barents Sea. See Dowdeswell & Scourse (1990), 269–88.

Wang Yuelen, Lu Songnian, Gao Zhenjia, Lin Weixing, Ma Guogan 1981. Sinian tillites of China. See Hambrey & Harland (1981), 386–401.

Webb, P-N., D. M. Harwood, B. C. McKelvey, J. H. Mercer, L. D. Stott 1984. Cenozoic marine sedimentation and ice volume variation on the East Antarctic craton. *Geology* **12**, 287–91.

Weertman, J. 1961. Mechanism for the formation of inner moraines found near the edge of cold ice caps and ice sheets. *Journal of Glaciology* **3**, 965–78.

Weertman, J. 1987. Impact of the International Glaciological Society on the development of glaciology and its future rôle. *Journal of Glaciology* (special issue commemorating fiftieth anniversary of the International Glaciological Society), 86–90.

Wentworth, C. K. 1922. A scale of grade and class terms of clastic sediments. *Journal of Geology* **30**, 377–90.

Wentworth, C. K. 1936. An analysis of the shape of glacial cobbles. *Journal of Sedimentary Petrology* **6**, 85–96.

Whalley, W. B. 1971. Observations of the drainage of an ice-dammed lake – Strupvatnet, Troms, Norway. *Norsk Geografisk Tidsskrift* **25**, 165–74.

Whalley, W. B. & D. H. Krinsley 1974. A scanning electron microscope study of surface textures of quartz grains from glacial environments. *Sedimentology* **21**, 87–105.

Wharton, R. A., B. C. Parker, G. M. Simmons, K. G. Seaburg, F. G. Love 1982. Biogenic calcite structures forming in Lake Fryxell, Antarctica. *Nature* **295**, 403–5.

Whillans, I. M., J. Bolzan, S. Shabtaie 1987. Velocity of ice streams B and C, Antarctica. *Journal of Geophysical Research* **92** (B9), 8895–902.

Williams, G. E. 1989. Late Precambrian tidal rhythmites in South Australia and the history of the Earth's rotation. *Journal of the Geological Society* **146**, 97–111.

Wilson, R. C. L. 1991. Sequence stratigraphy: an introduction. *Geoscientist* **1**, 13–23.

Woodworth-Lynas, C. M. T. & J. Y. Guigné 1990. Iceberg scours in the geological record: examples from glacial Lake Agassiz. See Dowdeswell & Scourse (1990), 217–34.

World Glacier Monitoring Service 1989. *World Glacier Inventory*. IAHS (ICSI)–UNEP–UNESCO.

Worsley, P. 1974. Recent "annual" moraine ridges at Austre Okstindbreen, North Norway. *Journal of Glaciology* **13**, 265–77.

Wright, R. & J. B. Anderson 1982. The importance of sediment gravity flow to sediment transport and sorting in glacial marine environment: Eastern Weddell Sea, Antarctica. *Geological Society of America, Bulletin* **93**, 957–63.

Yoon, S. H., S. K. Chough, J. Thiede, F. Werner 1991. Late Pleistocene sedimentation on the Norwegian continental slope between 67° and 71°N. *Marine Geology* **99**, 187–207.

Zilliacus, H. 1989. Genesis of De Geer moraines in Finland. *Sedimentary Geology* **62**, 309–17.

Zotikov, I. A. 1986. *The thermophysics of glaciers*. Dordrecht: Riedel.

Zotikov, I. A., V. S. Zagorodnov, J. V. Raikovsky 1980. Core drilling through the Ross Ice Shelf (Antarctica) confirmed basal freezing. *Science* **207**, 1463–5.

Index

293

294

295